DESIGNING HUMAN PRACTICES

DESIGNING HUMAN PRACTICES

An Experiment with Synthetic Biology

PAUL RABINOW AND GAYMON BENNETT

The University of Chicago Press   Chicago and London

Paul Rabinow is professor of anthropology at the University of California, Berkeley. He has written numerous books, including *Making PCR: A Story of Biotechnology* and *The Accompaniment: Assembling the Contemporary*, both published by the University of Chicago Press.

Gaymon Bennett is a senior research fellow at the Center for Biological Futures at the Fred Hutchinson Cancer Research Center. He is coauthor of *Sacred Cells?: Why Christians Should Support Stem Cell Research*.

The University of Chicago Press, Chicago 60637
The University of Chicago Press, Ltd., London
© 2012 by The University of Chicago
All rights reserved. Published 2012.
Printed in the United States of America

21 20 19 18 17 16 15 14 13 12   1 2 3 4 5

ISBN-13: 978-0-226-70313-8 (cloth)
ISBN-10: 0-226-70313-4 (cloth)
ISBN-13: 978-0-226-70314-5 (paper)
ISBN-10: 0-226-70314-2 (paper)

Library of Congress Cataloging-in-Publication Data

Rabinow, Paul.
    Designing human practices : an experiment with synthetic biology / Paul Rabinow and Gaymon Bennett.
        p. cm.
    Includes bibliographical references and index.
    ISBN-13: 978-0-226-70313-8 (cloth : alkaline paper)
    ISBN-10: 0-226-70313-4 (cloth : alkaline paper)
    ISBN-13: 978-0-226-70314-5 (paperback : alkaline paper)
    ISBN-10: 0-226-70314-2 (paperback : alkaline paper)
1. Synthetic biology. 2. Synthetic biology—Research—United States. 3. Synthetic biology—Moral and ethical aspects.
4. Bioethics. 5. Synthetic Biology Engineering Research Center.
I. Bennett, Gaymon, 1972–. II. Title.
    TA164.R33 2012
    660.6—dc23   2011040160

♾ This paper meets the requirements of ANSI/NISO
Z39.48-1992 (Permanence of Paper).

CONTENTS

ACKNOWLEDGMENTS

The design and research of our experiment in participant-observation as well as the book that chronicles its vicissitudes extended over a period of five years. The labor involved was arduous. The time of inquiry was filled with multiple instances of discordancies and indeterminacies, to use the technical vocabulary of John Dewey, which we found helpful throughout the experience. It is simply the case that we did not regularly encounter the warm glow of solidarity, curiosity, and reciprocity from those with whom we were attempting to collaborate. We came to understand, and even appreciate, that onerous experiences, which seemed all too often to be gratuitous given the stated willingness to collaborate, were themselves an integral part of the experiment itself, given its design and its goals. After all, the results of a rigorous experiment constitute objective facts and thereby contribute to a broader understanding of the contemporary situation in which the biosciences and the human sciences currently operate.

That being said, the intricate experiences of this experiment were illuminated, sustained, and encouraged by the care and insights of others. How fortunate we have been! Those others include authors such as John Dewey, Michel Foucault, Max Weber, Kant, Seneca, Aristotle, and others whose accompaniment through the medium of their books; their wisdom, work, and courage sustained and encouraged us on many a spiritually dark day here in sunny California.

Warmth and encouragement came from other quarters with a generosity, solidarity, reciprocity, curiosity, and care for which we are humbly grateful. First, we thank those who were closest to us during these experiments, especially Marilyn Rabinow and Christin Quissell. We wholeheartedly thank Anthony Stavrianakis, who, from the outset, collaborated tirelessly with us in the design of Human Practices. Deep thanks also go to friends and colleagues who provided continued intellectual engagement and support, especially James Faubion, Meg Stalcup, Timo Rodriquez, Adrian Van Allen, George Khushf, and Roger Brent.

We thank those at SynBERC and the NSF who helped us advance our work, especially Frederick Kronz, Kevin Costa, and Leonard Katz. And, of course, we could not have conducted our experiment without ongoing interactions with colleagues at SynBERC, particularly Jay Keasling, Drew Endy, Adam Arkin, Chris Anderson, George Church, Ron Weiss, Christopher Voigt, Wendell Lim, Clem Fortman, John Dueber, Ken Oye, and Julius Lucks.

The financial support for our work over the course of five years was the generous funding of the National Science Foundation, who sponsored SynBERC. In addition we had a seed grant from the Department of Energy.

Introduction: A Productive Experiment

> Reflection appears as the dominant trait of a situation when
> there is something seriously the matter, some trouble, due to
> active discordance, dissentience, conflict among factors of
> a prior non-intellectual experience.
>
> JOHN DEWEY[1]

> All knowledge, as issuing from reflection, is experimental.
>
> JOHN DEWEY[2]

This book provides an account of a productive experiment in the human sciences. We choose these terms knowingly; we are not using the term *experiment* in a metaphorical sense. We designed and attempted to implement a disciplined plan of convergent scientific and ethical engagement and inquiry. The design of the experiment built upon both authors' extensive experience in the study of the broad historical conditions under which genomics and post-genomics emerged, drawing on their prior work in anthropology, theology, bioethics, and philosophy.

Naturally, like any experiment—that is, any real experiment—unforeseen turns from what we anticipated occurred with some regularity; consequently, the experiment's original design parameters required periodic modifications. Thus, at times, the objects under study exhibited unexpected contours, opacities, and resistances; given these repeated happenings, our preliminary expectations for both the ontological and ethical parameters of the experiment had to be revised and adjusted to fit what we encountered during the course of the inquiry. Although each of these dimensions proved at times disconcerting or confusing, we gradually came to take up such deviations from the anticipated course of events as indicators that we were indeed conducting a true experiment. We learned a great deal during the course of carrying out our work. To the extent that we can formulate what we learned, by organizing it, reflecting on it, and making it available to enhance the design of further experiments (for ourselves and for others), we assert that we have been practicing science in the broad sense. And to the degree we have learned specific things with general significance, our experiment must be considered a success.

A main reason that we continued to learn and to incorporate insights along the rutted way was that we began the undertaking by developing and systematizing a repertoire of conceptual tools appropriate to the task and integrating them into our experimental design. By so doing, we were able to monitor the course of our experiences as well as to provide ourselves with a means of rectification (or at least reflection) on what we were observing. Many of these tools are available for scrutiny and use at our website: www.bios-technika.net.

Empirically, it is fair to say, however, that in the qualitative social sciences, it is unusual to claim (beyond the mandatory rhetorical prose required in grant proposals) that one is working in an explicitly experimental or at least hypothesis-driven mode. Indeed, reactions from colleagues have shown us that the term *experimental* can function as much as a hindrance to understanding as an aid. We have been asked: Are we being literal, figurative, analogical, metaphoric, or simply ironic? The answer is metaleptic, but that is another story for another audience.

We will use the term *inquiry* as our most general term of art to describe our undertaking. Following the diverse and rich investigations of John Dewey over his lengthy career, the term includes experimentation and experience, breakdown and repair, confusion and clarification, reflection and determinations—as integral elements of both the objects and the procedures of an imagined human science. Dewey called this practice "philosophy," but of course his use does not correspond in any tight fashion with the existing discipline of that name.

Ontology

The twin wings of our enterprise have been ontology and ethics. First, ontology: new things are made and introduced into the world, directly or indirectly, all the time—iPads, nuclear devices, venture capital firms, knock-out mice, modified nucleotides, gluten-free cinnamon waffles, designer jeans, *Wired*, Google, new varieties of apples and dogs and yeast, type 2 diabetes epidemics, the polymerase chain reaction, and social networking, to name a few contemporary ones. The issues of what things are, which new things are significant, how they are significant, and how one knows they are new and significant are of prime importance pragmatically, anthropologically, and philosophically. The term that Western philosophy has used to describe reflection on the essence of objects is ontology. Since the Greeks, philosophy has been concerned about what it means to name "the most general kinds of things that exist in the universe."[3] That challenge, like all important philosophical debates, has endured without definitive resolution for millennia.

We have not taken up the task of constructing a general ontology. Rather, what we propose is a contribution to an anthropology of the contemporary world. An anthropology of the contemporary world, among other things, adopts a strategy of designing techniques to change professional philosophic discussions into topics of inquiry in the broad sense of a rigorous, empirical investigation guided by conceptual reflection. How things come into existence—are named, sustained, distributed, and modified—is an issue of primary importance for many scientific disciplines, especially for synthetic biology, whose goal is precisely the creation of such objects. Observing, reflecting, analyzing, and representing these contemporary ontological changes has been a prime objective of our experiment.

One axis of the design of our experimental inquiry focused on a core claim made by synthetic biology's spokespersons that biology could now become an engineering discipline, and consequently new beings with different properties would soon populate laboratories and eventually a range of environments. We thought it vital to begin an inquiry into these potential (or virtual) entities. Given that we had privileged access to observe the efforts (biological and institutional) where this work was taking place, it seemed to us that it was scientifically more instructive to focus on what was happening in the laboratories and institutions rather than on metaphysical debates about Nature or God. Of course, we know full well that such metaphysical debates have other functions than the ones they announce, as well as ramifications that their practitioners may not intend. Those theological and bioethical debates (and the institutions and subjects authorized to conduct them) were, of course, themselves bringing things into the world. In the main, these things were discursive and demarcated from the coproduction of biological entities. In fact, the goal of much of this discourse was the prevention of certain biological entities (clones, engineered stem cells, designed genomes, etc.) that were imagined to be coming into being.

We situated ourselves in a different milieu with different aims. Where we did direct our inquiry to mainly discursive productions, it was in the founding manifestos of the small group of entrepreneurs who brought synthetic biology into existence as a thing-of-the-world. What those manifestos heralded and the processes through which the initial research agendas were forged in relation to them constitute an important object of our observations. This part of our experiment was relatively straightforward, if demanding, to carry out.

For a time we encountered resistance in the form of offhand derision about the term *ontology*, although eventually the mockers lost interest in what we were doing; those who remained attentive to the directions in which our inquiry was leading us eventually acknowledged that the gene ontologies that their colleagues in genetics and molecular biology were producing and posting on the Internet were not simply phantasmagoric or postmodern; perhaps our efforts at ontology weren't either.

In this instance, as in others, one major finding of our experiment was the existence of our SynBERC colleagues' wide-ranging lack of curiosity outside of their specialties. On the one hand, their indifference meant that we were unobstructed, even aided, in pursuing our research; on the other hand, when it came to the ethical problems, this indifference and lack of curiosity proved to be a different matter. It proved to be, to quote John Dewey again, a site where there was something "seriously the matter, some trouble, due to active discordance, dissentience, conflict among factors of a prior non-intellectual experience."[4] We will explore and discuss aspects of that prior non-intellectual experience at appropriate places throughout this book.

Ethics

The second wing of our undertaking has been ethics. Many bioscientists and bioethicists hold the view that the role of ethics is principally to restrict scientific excess. This perspective tends to presume that the most urgent problems can be known in advance of ongoing scientific practice and that concerns arise primarily from the use of technology. It follows that ethics must establish moral "bright lines" that science or medicine cannot cross. According to this understanding, bioethics serves to protect and defend the public, society, even humanity. We think that an approach to ethics based on these premises limits understanding of how past problems bear on contemporary situations and inhibits the identification of new problems. Furthermore, it undervalues the extent to which ethics can play a formative role in the very development of both biological and human science and technology.

Our approach builds on the lessons of past ethics programs while attempting to move beyond their acknowledged limitations. Like the development of bioethics in the late 1970s, we think it is crucial to assess the relationship between the near-future and present practices. Unlike the procedures put into place following the foundational Belmont report, we do not want to limit the intersection between ethics and science to the space of intermittent bureaucratic review. Like the Ethical, Legal, and Social Implications (ELSI) program of the Human Genome Sequencing Initiative), we believe it is valuable to think about how research shapes self-understanding. Unlike ELSI, we do not want to limit our work to reflection on downstream consequences of research. Finally, like the President's Council on Bioethics (PCB) when it was led by Leon Kass, we think it is of utmost important to explore the critical limits of science and human life. Unlike the PCB, however, we do not think that this work should consist primarily of establishing a priori moral boundaries.

We think that our contribution can only be effectively realized if ethical inquiry is conducted in direct collaboration with scientists, policy makers, and other stakeholders. This orientation is accomplished not through the prescription of moral codes, but through mutual reflection on the practices and relationships at work in scientific engagement and how these practices and relationships allow for the realization of specified ends. Straightforwardly, such an approach to ethics encouraged us to inquire into what was going on as it was being formulated, and to form initial evaluations of projects and strategies. Our goal has been to develop and sustain this mode of operation. Seen in this light, understood in this frame, both the problems addressed in our inquiry and the mode in which that inquiry was to be conducted had to be understood as ethical.

Discordant Modes: Cooperation versus Collaboration

Our contract with the National Science Foundation and the bioscientists and engineers at the Synthetic Biology Engineering Research Center (see chapter 1) was to invent a form of ethical practice that differed from the model invented and put into practice during the work of the Human Genome Sequencing Initiative. The governmental mandate in that project was for the ethicists to devote themselves to exploring the social, ethical, and legal consequences of the techno-scientific work undertaken to sequence the human genome. By contrast, our challenge was to invent a new way of working that was not simply downstream and outside of the techno-science but upstream and adjacent to this new domain of biological engineering.

A fundamental distinction exists between collaborative and cooperative modes of work. A cooperative mode of work consists in demarcated tasking on defined problems and objects, with occasional if regular exchange at the interfaces of those problems. Such cooperative exchanges are organized structurally from the outset. As a work mode, cooperation works best where problems and their principal features have been stabilized such that a cadre of experts and techniques can expand around those problems in a more or less rationalized manner. When such conditions are in place, the question of significance and ends can be taken for granted or held stable. The metric of such means-ends relations is efficiency and production. Cooperation, however, entails neither a common understanding of problems nor shared techniques of remediation. Cooperation assumes specialization and a defined division of labor.

A collaborative mode of work, by contrast, proceeds from an interdependent but not predetermined division of labor on shared problems. It entails a common definition of problems (or at least the acceptance of the existence of a problem-space) as well as an agreement to participate in the development and implementation of techniques of remediation. Collaboration is appropriate where problems and their significance are in question, or where the heterogeneity, complexity, instability, or yet to be determined dynamics of a problem-space require the interactively coordinated skills and contributions of co-laborers of diverse capacities and dispositions without knowing in advance how such work will be organized and what it will discover. Collaboration anticipates the likely reworking of existing modes of reasoning and intervention, adjusting these modes to the topography of the emerging problem-space. Collaboration proceeds with the assumption that new capacities, skills, arrangements, and distribution of power may well be required to carry out a successful inquiry.

The most significant difficulties in our experiment as well as our experiences turned on the attempt to imagine out what collaboration might look like and to turn those thoughts and reflections into practices. Traditionally, anthropologists have called themselves "participant-observers." The observation side has always been privileged, with the participation side of the oxymoron designating the fact that most anthropological fieldwork entails experiences away from the text in situations that they do not control for reasons generally opaque to those they are observing. Our experiment was one of participant-observation in a modified sense; we use the term *observation* (*Betrachtung*) here in the sense that the German sociologist Niklas Luhmann meant it—as both observation and intervention as a mode of inquiry.[5]

What did we want from our bioscientist and engineer colleagues? We wanted what we had formally agreed upon with the National Science Foundation in the first place—an experiment in upstream, collaborative practice. What would that look like? We did not know in advance—except that it would not consist primarily in us producing either public relations proclamations or direct policy initiatives. This starting point did not seem problematic: There is ample money and expertise available for both of these activities; there are existing governmental agencies and trade organization lobbyists and publicists galore trained and eager to perform such labor. Rather, we wanted to invent something more in the spirit of John Dewey's democratic process of inquiry in which common goals or problems are not defined in advance with any specificity but emerge through collaborative practice. We sought initiatives from the biologists and engineers about problems and concerns beyond their subspecialties as well as support from those in power, requiring the biologists and engineers to engage in open discussion and exploration.

Although promises were made and formalized, progress in the direction of collaboration was at best stuttering and sporadic. There was scant encouragement or engagement in the mutual production of new venues and forms in which and through which the production and modification of different kinds of scientific/ethical objects could be facilitated. In a sense, our experiment encountered and documented anthropologically what colleagues in the quantitative social sciences have reported on: cooperation among unequals as the acceptable compromise.[6] Scientifically, therefore, our experiment has yielded productive results. Ethically, however, significant discordancies remain. Consequently, other forms of collaboration with other instances will have to be undertaken.

Given that in emerging problem-spaces such as post-genomic biology current expertise is by definition insufficient to understand the new objects that are being produced and new experts do not yet exist, how to give form to collaboration remains a central challenge whose significance cannot be overestimated. Although this claim is readily accepted and supported when it comes to work internal to the life sciences, the same type of claim is rarely addressed seriously, and is easily neglected, when it comes to the inclusions of the human sciences in bioscientific enterprises. One of our most important conclusions is that such a situation is shortsighted and dangerous.[7]

Stakes: Amelioration, Prosperity, Flourishing

The dominant mode of rationality and purpose guiding the life sciences today is instrumental. This rationality is not new, and researchers and research directors have developed strategies for reconciling the vocational dimensions of scientific research with demands for instrumental outcomes.[8] However, diverse factors are contributing to an intensification of an instrumental orientation. These factors include: the predominance of the biotech industry as a widespread model for bio-scientific research; the demands of funding agencies (private and public) that experimental results be formulated so as to be on the road to commercialization; the mandate to equate the worth of science to instrumental norms as well as the dismissal of those who don't accept this equivalence; the express aspiration—whether simply performative or sincere—to make science serve some broader good; and, in the case of the experiment reported on in this book, a defining commitment on the part of those in synthetic biology and related domains, to remake biology through the adoption of an engineering disposition.

Through the use of anthropological techniques of participant-observation and analysis, we have identified two terms as candidates for naming the stakes of the enterprise (and, to a degree of their lives) currently prevalent among the biologists, engineers, NSF administrators, and industrial partners at SynBERC: prosperity and amelioration. All of the aforementioned players involved are unashamedly and unselfconsciously committed to prospering personally, institutionally, or nationally—most proceed as though all three are synergistically connected.

For the bioscientists and administrators (and their industrial partners) to prosper means primarily devoting dedicated and sustained attention to advancing their careers, which includes doing their job, whether scientific or administrative, well. It also means, in many cases, striving for financial success through involvement with start-up companies that they themselves have founded, as is generally the case for the more senior bioscientists, or wish to found in the case of the more junior ones. An official part of the mandate of the engineering centers funded by the National Science Foundation is to become financially self-sufficient within ten years. Given that mandate, it is assumed that there is no fundamental contradiction between what the government decrees and what the biologists and engineers desire: industry, scientific practice, and ethics are assumed to have a pre-established harmony.[9] By now the American scientific establishment accepts such practices and dispositions; in 1980 the Bayh-Dole Act mandated the potential for commercialization of all government-funded projects and has long since become not only the law of the land, but also an institutional norm.[10]

The engineers and biologists with whom we worked affirm, if asked, that it would be a good thing if their work contributed to an amelioration of the environment, health, and security. Pressing the issue further, we have discovered that few of them actually believe that attaining such amelioration is an immediate prospect, only something beckoning on an indeterminate horizon. They also affirm that the attainment of such ameliorative goals will have to be the work and responsibility of others—industrial partners, government agencies, or unspecified others. In a word, while maintaining a positive evaluation of amelioration, they take it for granted that it will be achieved through cooperation rather than collaboration.

None, to our knowledge, have expressed any deep concern that prosperity and amelioration might be conflicting or contradictory goals. Beyond that tacit and largely unexamined consensus, a zone of ambiguity exists. Are the metrics of prosperity and amelioration comprehensive and sufficient? We have found that posing such a question to the biologists and engineers yielded perplexity, indifference, and/or hostility.

A striking feature of the manifestos of synthetic biology has been precisely the proposition that through redesign and recomposition, living systems could be made to function differently, more efficiently, more precisely—that is, new capacities could be produced through the restylization of existing and newly invented materials.[11] Whether the maximization of functionality is good or bad has been bracketed as a legitimate question; as such, it is simply beyond the ken of synthetic biology.

Our efforts have been calibrated from the start to a different metric: flourishing. Flourishing as a metric involves more than success in achieving projects; it extends to the kind of human being one is personally, vocationally, and communally, as well as the venues within which such human flourishing is facilitated and given form as a practice. Throughout the history of Western civilization (at least before the nineteenth century and the rise of utilitarianism), philosophers, theologians, and ethicists have distinguished the good life (understood as *eudaimonia, summum bonum,* etc.) from technical optimization. In 2007 we produced a short manifesto on human practices, which turned on the importance of the concept of flourishing (see chapter 2). We introduce here the core stated in the manifesto:

> *Flourishing is a translation of a classical term,* eudaimonia, *and as such a range of other possible words could be used: thriving, the good life, happiness, fulfillment, felicity, abundance, and the like. Above all, flourishing should not be confused with technical optimization, as we hold that our capacities are not already known or fixed in advance. We do not understand flourishing to be uncontrolled growth, progressivism, or the undirected maximization of existing capacities. Here we are merely insisting that the question of what constitutes a good life today and the contribution of the biosciences to that form of life must be vigilantly posed and re-posed.*

Which norms are actually in play and how they function must be observed, chronicled, and evaluated in an ongoing fashion.

The core question and challenge of our experiment, then, was this: How might synthetic biology be made to contribute to (and participate in) a mode of practice guided by—if not uniquely dedicated to—an ethic of flourishing?[12] Our experiment as well as our experience turned on developing an adequate diagnosis of the current situation, of defining and delimiting a repertoire of concepts appropriate to that analytic task, of designing and contributing to a collaborative mode of practice—or, at the very least, of putting these challenges and their stakes on the contemporary agenda.

PART I

Human Practices: Diagnosis

1 The Setting

SynBERC: The Synthetic Biology Engineering Research Center

> How can the growth of capacities be disconnected from the intensification of power relations?
>
> MICHEL FOUCAULT[1]

In the wake of the various genome-sequencing projects of the 1990s, as well as the security sequels of 9/11, the globalization of information and finance capitalism, the life sciences have undertaken the challenge of redesign and coordination.[2] It is a central working hypothesis of this book that the life sciences and the human sciences, as well as the relations between them, are currently unresolved about the status of their objects, the best venues in which to work on them, and the broader ethical framing of their undertakings.

If today there is a broad consensus that the genome sequences were not the key to life, only the "end of the beginning" of biology, as Sydney Brenner put it, then it follows logically at least that the ELSI (Ethical, Legal, and Social Implications) programs—which were constructed within that earlier political and scientific consensus about the significance of the genome-sequencing projects, while continuing to provide useful safeguards and as venues for conducting public conversations—are themselves limited in their scope by their original mandate to operate downstream and outside of the sequencing efforts.[3]

Agreeing with Brenner that there is a compelling need for scientists to rethink current understandings of *the gene*, we argue in a parallel fashion that there is an equally if not more compelling need to rethink the cornerstone concept of ELSI—*social consequences*. The need for rethinking what is meant by *social consequences* is actually more compelling because while it is habitual for normal bioscience that outdated concepts will sooner or later be replaced, there is no guarantee whatsoever that a parallel process exists for the human sciences (not to mention media discourse). It follows logically—although many pragmatic obstacles to making it a reality remain in place—that contemporary post-genomic research programs such as synthetic biology can no longer be constituted as they were in the recent past.[4] We argue that the challenges of rethinking the ramifications taking place within the biosciences and the human sciences should be addressed. How to make this twofold task collaborative and synergistic, however, remains problematic.

Diagnosis: The Recent Past

After the completion of the human genome-sequencing projects, it became clear to most observers (and many participants) that the nucleotide sequences themselves were neither the "holy grail" nor the "code of codes" that the proponents of the projects hoped they would be. Nor were these seemingly endless strings of base pairs the key to "playing God" or "Franken-futures," as opponents warned. By the early years of the twenty-first century, whatever work these analogies had originally been designed to do, they had become outmoded and misleading. It is now clear that the sequence information is one of the most important foundational elements—necessary but hardly sufficient—for constructing a contemporary biology.[5]

What was missing most conspicuously was a credible scientific program for moving from the hope (and desire) that bio-informatics would provide the technological means to deciphering an ever-increasing quantity of molecular information to a more closely calibrated strategy for laboratory experimentation in the near future. Correlatively, an honest inspection revealed an even bigger gap between the overflow of information and its promised transformation into ameliorative and lucrative applications. Finally, there was an amorphous but haunting awareness that what was required ultimately was a firmer scientific understanding of the material under consideration, an explanatory frame adequate to biological structure and function beyond suggestive statistical correlations and broad generalizations about life.

This overabundance of data and under-determination of its significance have yielded a surfeit of visions-*cum*-manifestos. The manifestos have been driven by the need to articulate and defend a new mission for the large bureaucracies and their costly technologies and facilities that had been constructed as part of the sequencing projects; by a desire to attract venture capitalists and other significant funders; by a drive to develop and implement research strategies that would be scientifically and financially rewarding; and so on. The hectic activity devoted to defining the framing and analogical correlatives of a convincing post-sequencing orientation goes some way to situating the effervescent (and largely evanescent) efforts to brand and promote proteomics, systems biology, synthetic biology, and the like, as the crucial next stage in bringing into existence the hoped-for wonder and bounty of a biologically based future of knowledge, health, and wealth, which had been so forcefully articulated and promoted by the proponents of the sequencing projects.

Equally significantly, but with less hoopla, by 2007 the ethics initiatives that had come into existence as part of the sequencing projects—the ELSI (Ethical, Legal, and Social Implications) programs—were also beginning to be critically scrutinized.[6] These programs were constituted according to the terms of a political agreement among the Human Genome Project funders that ELSI would be supported on condition that it operated downstream of the science and technology, and should concern itself primarily with framing *social consequences*. The demand for rethinking this approach has come in part initially from the governmental funders of the multiple centers in nanotechnology and then in synthetic biology—that is, the U.S. Congress and the National Science Foundation.

Inquiry: The Near Future

One exemplary area of the new post-genomic life sciences is synthetic biology. Starting out as a placeholder term, or perhaps a hoped-for brand, during 2007 *synthetic biology* began to coalesce into a number of defined research programs. In its early years, synthetic biology has received attention from media and funders for two principal reasons: There are the audacious claims made by some spokespersons that synthetic biology will fashion living systems into—pick your analogy—biological LEGO sets, plug-and-play genetic robots, or genetic circuits. We have been told that (a) biological complexity will be refactored into simple constituent parts, and rational design and composition made child's play (that is to say, undergraduates and high school students will be doing it with increasing facility); (b) the capacity to design and manipulate biological systems is uniquely suited to solve many of the world's most pressing problems.[7] The self-styled prophets confidently assert that synthetic biology is going to discover new therapeutics and lower their cost, afford the means to solve the energy crisis, be the key to biosecurity, and repair the environment.

Most broadly, post-genomics has seen the intensification of an engineering disposition in biology: understanding through making and remaking. Living systems, and their components, are being redesigned and refashioned. The challenge for synthetic biologists is to take biology beyond the guild-like restrictions of artisanal *savior faire* and to make it into a full-fledged engineering discipline, with all this entails in terms of standardization, modularization, and regularization. The task is to design and fabricate useful biological objects, with an express commitment to understanding only enough about how they work to make them work. Though there is disagreement about how exactly this feat might be accomplished, there is agreement that the goal of standardized biological engineering will require a re-assemblage of scientific subdisciplines, diverse forms of funding, institutional networks, governmental and nongovernmental agencies, legal standards, and the like.

In 2006 the National Science Foundation made a significant investment in synthetic biology through its Engineering Research Centers (ERC) directorate. The funding of the Synthetic Biology Engineering Research Center (SynBERC) was conditional on the inclusion of an equal and integrated "social implications" component. Our charge has been to fulfill this mandated condition. We have asked: What should that component look like? In the fall of 2006, one of us (Rabinow) traveled to Washington, D.C., to meet with the ERC directorate officials: to their credit, the officers at the NSF were frank in admitting that they did not know and encouraged innovative exploration of what such a component might be. Hence a challenge was put into play in a fashion parallel to the challenge of constituting a program for post-sequencing biology: What form should be given to a biological engineering center that incorporates collaboration as equals with human scientists?

What follows is an account of conceptual and ethical diagnosis and inquiry, as well as an attempt to organize, orient, and evaluate the human practices work that we agreed to undertake as one of SynBERC's four "thrusts" (Parts, Devices, Chassis, Human Practices). The chapters of this book were written over the course of four years (2006–10). Although some minor adjustments have been made to the texts to minimize redundancy, we have explicitly resisted rectifying all of our prior formulations so as to preserve the temporality and context of these interventions. In this book, our purpose is to provide a kind of archive and chronicle of shifts in temperament, expectation, conceptualization, as well as scientific and ethical capacity—or its neglect. At each of these junctures in our experiment, we faced the challenge of assessing, in real time, how our experiment was ramifying and how best to proceed.

The Site: SynBERC

During 2006, a group of researchers and engineers from an array of scientific disciplines proposed a project with the aim of rendering synthetic biology a full-fledged engineering discipline. Representing five major research universities—University of California, Berkeley; Massachusetts Institute of Technology; Harvard; University of California, San Francisco; and Prairie View Agricultural and Mining—the participants proposed to coordinate their research efforts through the development of a collaborative research center: the Synthetic Biology Engineering Research Center, or SynBERC (www.synberc.org). SynBERC is highly unusual on a number of counts. In addition to its far-reaching research and technology objectives, it embodied an innovative assemblage of multiple scientific subdisciplines, diverse forms of funding, complex institutional collaborations, an orientation to the near future, and intensive work with governmental and nongovernmental agencies, focused legal innovation, and imaginative use of media.

The reviewers and officials of the NSF reacted enthusiastically to the ambitious proposal. Before making the official award, however, NSF officials informed Jay Keasling, a professor of chemistry at UC Berkeley and the future director of the Center, that the award was contingent on including a "social implications" component. In the wake of the events of September 11, 2001, the proposal's dual goal of (1) making biology easier to engineer and of (2) making materials and know-how openly available raised concerns about potential security ramifications. Keasling and his colleagues were perfectly willing to accept the need to address these issues, although neither the NSF nor the principal scientists and engineers who were to guide the Center had a well-formulated idea about what such a "social implications" component would look like or what it would do.

Keasling (presumably) turned to the dean of public policy at Berkeley for advice (or was approached by the dean). The dean proposed that an adjunct professor, Stephen Maurer, a lawyer with strong interests in economics, would be a suitable person to lead this component. Keasling, following the informal style of leadership that characterizes his approach to such matters, accepted the proposal. The short tenure of the first occupant of this ethics position was a troubled one. Maurer proposed, and argued forcefully for, two things. First, he advocated a mechanism to monitor "experiments of concern." Second, he underlined the need for a procedure whereby the "community" of synthetic biologists would vote on a set of regulatory controls that would govern the relations of the nascent DNA synthesis industry and the community of synthetic biologists. The substance of Maurer's proposals was eventually worked out in a report funded by the Sloan Foundation.[8] His proposals consisted in drawing attention to the need to monitor the solicitation of DNA sequences that could be identified (by checking against an as yet to be developed database) as of possible use in known pathogenic agents. Although most of the concerned actors took the substance of Maurer's proposals to be reasonable and desirable, personality conflicts and a battle over who was to set the terms for governance and potential regulation built to a point of total breakdown. After a contentious

behind-the-scenes set of confrontations, in June 2006 Keasling and his colleagues pulled Maurer's proposals at the very last minute from the agenda
of Synthetic Biology 2.0, at UC Berkeley. This was done without informing
Maurer, and no vote was taken.

In the wake of this theatrical turn of events—one that foreshadowed
a governance style and a use of unequal power relations that lingers at
SynBERC—Keasling invited Paul Rabinow, a professor of anthropology at
Berkeley, and Ken Oye, an associate professor of political science at MIT, to
jointly direct the so-called ethics, social consequences, public perception,
legal considerations, risk assessment, and policy implications component.
Both had been speakers at SB 2.0; each had found the other's presentation
interesting. The proposal made sense as there appeared to be a clear division
of labor, with Oye concentrating on policy issues and Rabinow on ethics
and the innovations in organizational form as well as the scientific objects
to be produced by the Center. Oye and Rabinow accepted this formal arrangement and became the co-principal investigators of Thrust 4: Human
Practices.

Human Practices

The unexpected invitation to become active participants in the construction of a multidisciplinary Center was both welcome and enticing. It was
welcome in that Rabinow had already contributed to the early developments in synthetic biology, albeit as an anthropological observer.[9] He had
been asked to give a presentation at the first international conference on
synthetic biology, SB 1.0, at MIT in 2005 and another at SB 2.0 at Berkeley
in 2006.[10] It was enticing given the programmatic statements that characterized the Center's initial strategic plan. One of us (Rabinow) proposed to
the other (Bennett) that we take up the challenge together. Bennett, at the
time a student at the Graduate Theological Union in Berkeley, had been attending seminars in the anthropology department. Hence our collaboration began with two years of salient conceptual work already under way.

We agreed that it would be an exciting challenge to try to think through
and put into practice a form of "post-ELSI" program. The mandated Ethical,
Legal, and Social Implications program of the Human Genome Sequencing
Initiative, while valuable in a number of ways, could not serve as a direct
model for the future. Essentially, the ELSI model (to simplify but not be-
tray) had a mandate to work outside and downstream of the technological
and scientific work. ELSI's directive was to deal with consequences, specifically "social consequences." There was a broad agreement that at SynBERC

(as well as at the NSF-funded nanotechnology engineering centers) the ethics work should be conducted alongside and collaboratively with the engineering programs.

We were fully aware that the power relations between the life sciences and human sciences were certain to be unequal. For more than a decade, one of us (Rabinow) had conducted anthropological work in the worlds of biotechnology and genomics. One of us (Bennett) had spent several years engaged as a bioethicist working on genomics and stem cell research. Hence, we were both aware that ambitious life scientists would have had a minimum of preparation and education, not to mention even an awareness, of the issues and developments in the human sciences and ethics in recent decades. We were aware that government officials might well be well-meaning but that they were under pressure to produce "first-order deliverables" and that their openness was likely to fade as pressures on them mounted from within their own institutions to have such deliverables. Nonetheless, against fairly large negative chances of success, the time seemed ripe to take a proverbial plunge and to see whether one form or another of collaboration could be designed (and put into practice).

After all, SynBERC was designed, proposed, and funded as an effort to invent new venues and research strategies capable of producing resourceful solutions to real-world problems where existing venues and strategies appeared to be insufficient. As their website puts it with the typical bravado of an early stage undertaking:

> The richness and versatility of biological systems make them ideally suited to solve some of the world's most significant challenges, such as converting cheap, renewable resources into energy-rich molecules; producing high-quality, inexpensive drugs to fight disease; detecting and destroying chemical or biological agents; and remediating polluted sites.

This undertaking was designed around four core research "thrusts." Research Thrusts 1–3 (Parts, Devices, Chassis) were to focus on the key technical challenges. For its part, Thrust 4 would work to develop synthetic biology within a frame of human practices. We proposed the name Human Practices to differentiate the goals and strategies of this component from previous attempts to bring "science and society" together into one frame so as to anticipate and ameliorate science's "social consequences." The task of Human Practices, thus, was designed as an effort to pose and re-pose the question of the ways in which synthetic biology is contributing or failing to contribute to the promised near future through its eventual input into medicine, security, energy, and the environment, as well as the ways in which a new generation of young bioscientists were being trained into the practice of science.

SynBERC Governance

Despite the formal inclusion of Thrust 4 as an equal component of the overall enterprise, active resistance to our research priorities on the part of the SynBERC Industrial Advisory Board (IAB) and some NSF divisional officials, as well as the persistent asymmetry of power between the human and biological sciences, made it difficult, and at times impossible, to meet our stated goals. Support for our work was not entirely lacking. Indeed, during the first three years, our approach and experiment were strongly supported by select members of the NSF's annual site review teams, as well as by several members of SynBERC's Scientific Advisory Board (SAB). In principle, the site review teams and the SAB were central to the governance of SynBERC, and, for a time, support from these quarters sufficed for protecting and encouraging our work. By year three, however, the NSF effectively backgrounded SAB in favor of the IAB in the governance of the Center.

The SynBERC Administration

At the first two annual site reviews, Jay Keasling proudly reported that the administrative costs of SynBERC were exceptionally low. One of the reasons for this low expenditure was the skeletal character of the administrative structure. Until midway through the Center's second year when Leonard Katz was hired as the scientific director and industry liaison officer, Kevin Costa, the Center's administrative director, was essentially the sole person authorized to coordinate the Center's activities.

During 2006, the first year of SynBERC's operations, Keasling was named *Discover* magazine's "Scientist of the Year" for his work on the malaria drug precursor artemisinin; during 2007 and 2008, Keasling was a chief architect of two major biofuels projects at UC Berkeley (totaling several hundred million dollars), as well as a founder of Amyris. In 2009 Keasling was named the CEO and director of the Joint BioEnergy Institute (JBEI) as well as the interim deputy director of the Lawrence Berkeley National Laboratory. Keasling had become a globally prominent scientist, with all this entails in terms of high-level engagements with other researchers, government bureaucrats, venture capitalists, and the like—engagements requiring near-constant travel. Keasling's rise meant that the SynBERC administrative staff was put in the untenable position of trying to develop governance structures and working scientific relations among and between the research programs of the sixteen SynBERC PIs from across the five member institutions.

Keasling's proposal had been attractive to the ERC directorate (and hence fundable) for at least two reasons. The first reason was simply that Keasling had recruited many prominent figures in bioengineering to serve as the Center's principal investigators (PIs). Of these PIs, the leaders of the four research thrusts during the first two years illustrate the point. The leader of the Parts thrust was UCSF microbiologist Wendell Lim. Lim's work on the evolutionary specificity of protein interactions was cited by some as justification for the feasibility of a parts-based approach to engineering biological systems. During SynBERC's first two years, Lim was selected as a Howard Hughes Investigator and became head of another major research initiative funded by the NSF studying cell propulsion. The leader of the Devices thrust was MIT engineer Drew Endy. Although Drew Endy's research portfolio was not of the stature of Lim's or the other thrust leaders, he was the most prominent spokesperson for synthetic biology. Without Endy's vision for synthetic biology and his contacts in Washington, D.C., it is unlikely SynBERC would have been funded. The leader of the Chassis thrust was Harvard Medical School genomicist George Church. Church, a major figure in the Human Genome Project, was (and still is) considered to be one of the elite biologists worldwide working on whole-genome engineering. Rabinow was the scientific leader of the Human Practices thrust.

Effective administration of a geographically, institutionally, and disciplinarily distributed center such as SynBERC would be difficult under any conditions. It was complicated by the status of the SynBERC PIs, which had both positive and negative effects. The scientific prestige and reputation of the PIs served as a formidable guarantee of the Center's credibility. And although several of the PIs did not actively participate in the life of the Center, their contributions to the annual reports and their presence at the site reviews provided a measure of reassurance for the NSF officers that SynBERC was positioned to meet its stated scientific goals.

This status and credibility, however, made it difficult for Keasling to make strong demands of the participating PIs, had he wished to do so. Given that these researchers were invited to participate in the Center on the basis of their ongoing research programs, it is hardly surprising that the SynBERC administration had a difficult time getting the PIs to restructure their research priorities and initiatives to better align with SynBERC's core agenda. For example, a central long-term deliverable promised by Keasling and colleagues was the formulation and testing of a first generation of

standards for parts-based engineering—standards for physically connecting individual parts as well as for ensuring their function when composed.
Within and among the PIs, Keasling made it clear that this goal would be
met through reporting on work and informal exchange. No formal mandate was given to the PIs that they either had to propose or conform to a
given set of standards. Moreover, a proposal made by the Human Practices
thrust that SynBERC create a "standards lab" to receive, review, parse, and
propose standardization of the PIs work was rejected as un-enforceable.

The profile and status of the SynBERC PIs—to repeat the point—simultaneously issued in credibility and strategic advantage as well as management difficulties and lack of scientific integration. This situation was further complicated by the NSF's expectations about how work within and
across the SynBERC labs was to be conceptualized and (hopefully) thereby
scientifically and operationally integrated. Keasling had proposed a coherent vision for the scientific organization of the Center. PIs would be divided
among the four research thrusts and three test-beds. PIs working in Thrust
1 would provide parts for people building devices in Thrust 2, who would
construct these devices to work in the chassis being designed in Thrust 3.
Specifics across these three thrusts would be determined by the needs of the
three test-beds, which would thus function to test the coherence and oper-
ability of the constructs being made, would provide the specifications for
work in the thrusts, and would also, crucially, create technologies of interest to industry. Human Practices, it was proposed, would be an integrating
factor throughout, connecting work in the thrusts and test-beds by way of
questions of security, intellectual property, ontology, and ethics.

The problem was simply that Keasling had envisioned the management
of the Center as composed of equals. The PIs were Keasling's colleagues,
and this was how he was going to treat them. In this light, Keasling took the
position that SynBERC was a democratic organization that should be run by
the principal investigators. To this end, Keasling had his administrative staff
coordinate a series of monthly conference calls during which PIs would take
it in turns to report on their research and discuss its relation to SynBERC
goals. The conference calls did not, in the end, transform the distributed
PIs into an effective deliberative body, despite several experiments with format, timing, number of participants, and the like. Two years in, Keasling
narrowed the governance of the Center to a representative committee of
PIs, consisting of the thrust leaders and a few others. In year three, this committee was narrowed yet again to a small management team of select PIs,
which simply reported its decisions to the other PIs during intermittently
scheduled conference calls. Even with this narrowed governance structure,
strong demands were rarely made of the PIs in terms of their scientific pri-

orities and deliverables. The human practices component was the exception, a point of considerable consequence, as will become clear.

At the insistence of the NSF ERC directorate, this programmatic view of the relationships among and between the various domains of SynBERC's effort was further rationalized according to a "three-plane diagram." An invention of Lynn Preston, the ERC director, the three-plane diagram (which was to be used by all the ERCs) consisted in a "knowledge base," a "technology base," and "technology integration." The SynBERC administrators were told that they needed to map the Center's four research thrusts and three test-beds onto this three-plane diagram, and to designate how operational flows would allow work to circulate up the planes and into the real world.

Within a very short period of time, practical difficulties with integrating work among and between parts, devices, chassis, and the test-beds were readily apparent. The divergence between the coherence of Keasling's proposal, the three-plane diagram, and the actual workings of the Center were striking. This gap became something of a joke among SynBERC PIs. These tensions have not, however, proven especially consequential. At least they have not resulted in the defunding of the Center, despite yearly threats to do so at the site reviews.

The lack of sufficient scientific integration and the failure of Keasling to take a sufficiently strong hand in directing the Center were noted in the year three site visit report:

> There is a significant threat to the future strength of the Center through its current approach to governance and management. The ultimate responsibility for project section and allocation of resources should reside with the Director. It appears that these decisions are made by consensus among the PIs and without consideration of input from both the IAB and SAB. This poses the risk of undermining the role of the Director in the governance and the mission of the Center. SynBERC also maintains a unique policy of a vote of confidence in the Center Director every two years. This policy weakens the authority of the Center Director, undermines his ability to make difficult decisions, and is contrary to ERC governance policy as indicated in the cooperative agreement. In addition, as reflected on the organization chart of the Center, the overall management system is confusing and must be revised to allow a more effective management with clear lines of authority and responsibility.[11]

The SynBERC executive committee never directly responded to this diagnosis either in terms of significant management reorganization or in terms of Keasling choosing to exercise his authority to make or enforce hard decisions.

The NSF ERC Directorate

SynBERC is funded by and falls under the governance of the NSF's Engineering Research Center directorate. Over the course of three decades, the NSF established, under congressional mandate, several dozen ERCs for work focused on the engineering of complex systems, with the explicit and ambitious goal of generating new platforms for industrial innovation. SynBERC was created and funded in this framework. SynBERC is one of five active ERCs funded with the aim of advancing biotechnology, and the only ERC with a focus on synthetic biology. Its administrators are required to attend national meetings for the administrative directors of the ERCs. Hence, there is a national and comparative perspective within SynBERC that provides a history of experience and expectations for how ERCs should be organized and administered and what counts as worthwhile deliverables.

During the first three years of the Center, the officials in charge of SynBERC were Lynn Preston (the ERC director) and Sohi Rastegar, the ERC co-director (in year four Rastegar accepted a new appointment, taking him out of the direct management of the Center). These officials were clearly charitable toward the governance and engineering achievements of SynBERC, as were their expectations for SynBERC's integration and productivity. They were also clearly on a learning curve of their own concerning the technical details of the biological research program.

This was also the period during which Jay Keasling was achieving national and international prominence. With the ascension of his Lawrence Berkeley National Lab colleague Steven Chu as the secretary of energy in the Obama administration, additional stimulus funds flowed west.[12] Keasling's credibility in Washington, D.C., and beyond was high. There was no doubt that these NSF officials were willing to take Keasling at his word when he promised that SynBERC would realize and surpass its stated ambitions.

One of the chief governance mechanisms that the ERC directorate uses to manage its programs is an annual site visit review and report. Each December SynBERC administrators and PIs assemble a massive (700+ pages) report on the Center's activities for the previous year. This report is intended to provide background and orientation for a team of reviewers. Two months after submission of the report, this team comes to Berkeley to conduct a forty-eight-hour review of the Center's research and progress. This formal site visit is preceded by a two-day all-hands retreat, which is essentially a dress rehearsal of the presentations that will be given at the site visit.

The purpose of the site visit and the preparatory report is both evaluative and prospective. This means that not only does the site visit team audit past activities, but they also provide an assessment. The assessment serves both as a mechanism for reorienting the Center's research and strategic priorities and, equally importantly, to provide criteria according to which the next year's report and site visit can be justly conducted. A key element in these procedures is the production of a site visit report. This report, which serves as the NSF's official statement on the status and value of the ERC, is composed after SWOT (Strengths, Weaknesses, Opportunities, and Threats) exercises and response by the PIs. The presentations, SWOT, and report are compiled over the course of the forty-eight-hour review. Although this review process is compressed and tiring, the first three years' reviews were largely supportive of SynBERC generally and the work of the Human Practices thrust in particular.

From the first site review forward, there were evident confusions about and reluctance to understand the mode and scope of the Human Practices strategy and research agenda. Such reluctance was expressed both by the engineers and biologists on the site review team, as well as certain of the NSF officials. However, given in years two and three the presence of two site visit team members conversant with the social studies of science and bioethics, the overall evaluation of Human Practices remained supportive and enthusiastic. This assessment was congruent with that of the bioscientific components in its willingness to support the emergent stages of this new enterprise, while also making it clear that this initial experimental period was expected to issue in a model of human science collaboration for other centers as well as in a series of familiar concrete deliverables, such as whitepapers and policy recommendations.

The site visits typically occurred at the end of February. The site visit team's report was circulated by mid-March. Over the ensuing weeks, the PIs were expected to compose a response, which Keasling oversaw, coordinated, and communicated to the NSF officials. In principle, the response was supposed to consist in strategies for redesigning SynBERC's research portfolio in light of the specific strengths and weaknesses identified by the review team. Subsequent to this response, in turn, the PIs were required to submit proposals for the following year's full-time equivalency allotments, congruent with these suggested strategic reorientations. In reality, the structures and rhythms of the academic system were completely out of sync with this schedule; for example, graduate students—the principal source of labor in all of these projects—were recruited and funded on yet another calendar. Consequently, this structure effectively meant that major changes in research priorities were unlikely to be implemented in such a short period of time, if at all.

The Industrial Advisory Board (IAB)

All NSF Engineering Research Centers (ERC) are mandated by congressional legislation to integrate industrial goals and partnerships into the running and financial maintenance of the ERC. These partnerships are legislated to follow an arc from minimal involvement and financing in the initial years through paid membership on the IAB, to total financing of the center by year ten through the creation of intellectual property (IP) and the direct transfer of valuable technologies into industry. For the first three years, the IAB at SynBERC, while present, essentially contributed advice and did not yet play a significant role in the governance of the Center.

Thus, during the first two years (2007–8), the visibility and operational significance of the IAB remained low and somewhat inchoate. They did not operate with a single voice, and most of their comments circulated as individual opinions expressed informally or during annual SWOT sessions, which at the end of the Center's annual site reviews served as a prelude to the official questions to which the SynBERC PIs were required to respond.

The informality and advisory role of the IAB in the initial two years is exemplified by two episodes: the place of "open source" in synthetic biology and the worth of the specific SynBERC test-bed projects, specifically the Tumor Killing Bacterium project. From the outset, PIs from MIT vigorously promoted an open-source approach to synthetic biology as a central pillar of their manifestos, both within and outside of SynBERC. Hence, the Bio-Bricks Foundation in Cambridge, Massachusetts—the key spokespersons and architects of open-source strategies for synthetic biology—included several SynBERC PIs on their board. During the first two years at SynBERC, possible strategies for implementing open-source and IP issues in general were frequent topics of formal presentations and discussions at the annual retreats and site reviews. Its advocates sought to balance their commitment to open-source biology with their active efforts to found companies. Although it was clear that IAB members were not enthusiastic about this discourse on openness and synthetic biology, and while Leonard Katz, the industry liaison officer, was contemptuous and dismissive of the critique of dominant property rights practices, the issue was not forced.

At the SWOT session of the second-year site review (2008), a number of IAB members forcefully and definitively dismissed one of the two test-beds, Chris Anderson's project to engineer bacterium to identify and destroy cancer cells *in vivo*. An illustration of one of Anderson's engineered cells was on the official poster for the event. Members of the IAB dismissed Anderson's work on the grounds that he had insufficiently considered and adopted the standard industrial models for cancer research. In fact, he had considered them but had decided to pursue a different strategy. Not only was the IAB attack on Anderson's work aggressive, but it was leveled at the most junior of the SynBERC PIs and was the only substantive complaint the IAB had about SynBERC's research portfolio. Although ultimately nothing came of this attack, this was the first indication that the IAB was beginning to flex its muscles. Only the Human Practices PI defended Anderson and his work, pointing out that industrial models and standards in cancer research had not demonstrated major curative triumphs, and that perhaps it was time to try alternative approaches. Keasling could have exercised power and intervened but did not.

By the third-year site review (2009), the IAB's participation and position, as well as their financial contributions, were substantially consolidated, and the NSF afforded them a formal role in deliberations. For example, Todd Peterson, the chair of the IAB and an executive at Life Technologies, served as a session moderator, and members of the IAB actively participated in the NSF Site Visit Team's (SVT) questioning of the research presentations. More importantly, the IAB had closed-door meetings with the SVT during which the industrial members articulated expectations and priorities for SynBERC researchers.

During these first three years, the criticisms of Human Practices centered on conflicts between and problems with integrating the research agendas of the two co-PIs. There were also significant differentials in scholarly productivity. By year three, these problems and differentials had basically been resolved by Keasling's decision to make Rabinow the sole director of Thrust 4. In view of these changes in leadership, both the SVT as well as the IAB expressed enthusiasm and support for the prospects of Human Practices work.

In the site review report, the Human Practices PIs were asked to create a formal code of conduct with regard to issues of safety and to help provide clarity on matters of intellectual property, a task eagerly taken up by Oye and consistent with his remit. The overall assessment of Thrust 4 in the year-three site review highlighted the worth and challenges of our experimental undertaking:

> The challenges integral to Thrust 4 are in many ways similar to those of Syn-BERC more generally. In both cases, they are attempting to create new disciplines. Thrust 4 seeks to create a new kind of research culture where scientists and engineers are working with scholars in the humanities and social sciences to proactively identify and address ethical and social issues integral to their own research. Thrust 4 is a project to bridge Snow's two cultures, and in doing so, to cultivate a new ethos of responsible scientific and engineering practice. The SVT

acknowledges the difficulties integral to this challenge, but also seeks to high-light the reasons motivating such work and the opportunities associated with it. As SynBERC advances its goal of making biology easier to design, it advances a qualitative transition in human capacity to both understand and alter living systems. In doing this, it creates possibilities of action that run ahead of any social consensus that might inform the policies that, in turn, sustain existing conditions for flourishing communities. This is both an opportunity for even greater flourishing and a threat to existing modes of well-being.[13]

There were only two substantial negative evaluations of SynBERC's Human Practices efforts. The first concerned the need to better integrate the research portfolios of the Berkeley and MIT groups. The second underscored the lack of initiative being taken by the biologists and engineers to engage in collaboration. Indeed, from the outset, Sohi Rastegar, the NSF ERC directorate's co-director, championed the need to address the biologists' passivity and indifference in regard to issues of security, preparedness, ethics, and the other mandates of the Human Practices thrust.

Blockage

Several trends, whose significance would become apparent later on, were taking shape from the outset. During the first year, our fledgling attempts at forging a post-ELSI Human Practices thrust were met by the bioscientists, engineers, and social scientists either with a benign disinterest or were disrupted (or dismissed) by a self-assured demand to justify ourselves, rhetorically and institutionally. The bioscientists and engineers were not averse to including ethics as part of the enterprise, at least from their perspective. Synthetic biology was frequently framed by its chief spokespersons as more than a set of technological challenges. It was also framed as "perfectly suited" to solving "the world's most significant problems." Such framing is not new, either rhetorically or in terms of work modes. Established research habits, as well as the reward and career structures connected to them, have long included a cooperative interface with ethics. In one sense, then, the "we don't understand what you are saying" responses were perfectly legitimate and not surprising. Our Human Practices undertaking was a new one entailing a certain amount of muddling and testing of preliminary approaches and modes of presentation.

But over the ensuing months, it became clearer that more than an initial strangeness was involved. There was basically no effort made to do any of the background work that was required to make sense of some of our technical or scholarly terms. Still less effort was spent actively contributing to the work of forging collaboration. Such collaboration would require a change of work habits; and the proposal of such change was received, at best, as a burden to funding and career trajectories. These academically successful biologists and engineers by and large had been trained in the highly specialized American system and were rarely informed, or interested, in a broader range of topics and issues; especially if such interest required more than simply voicing opinions on matters of security and the like.

This asymmetry and conceit were familiar. After a decade or more working with people in the biosciences, one of us (Rabinow) knew that their general formation was restricted and that they were very unlikely to be either aware or troubled by this state of affairs. Having worked as a bioethicist in Washington, D.C., and Silicon Valley, one of us (Bennett) understood the structural positioning of ethics as either downstream and regulative, or outside and advisory. What was new in the SynBERC setting, however, was that the work called for was neither just an anthropological observation nor a bioethical consultation. Human Practices had a mandate to accomplish a certain program of collaborative work. Despite that mandate, there was an often polite but unbending refusal to make this engagement mutual— it seemed to be taken for granted as natural that members of the Berkeley Thrust 4 team were conversant with the molecular biology and eager to learn more of the chemistry and engineering. No reciprocity emerged nor was it encouraged (or discouraged) by the other PIs; it simply was not considered.

What remained therefore was a hierarchy of power and privilege. Despite a series of non-exchanges, we concluded that this exercise of power relations was not intended to stifle us but only to keep us at a distance. Many of the SynBERC principals, including the MIT Human Practices group, tacitly assumed that the ELSI mode of external and social consequences was the norm, and a perfectly good one at that. In contrast, a post-ELSI undertaking required a change in habits, dispositions, and expectations during the process of forming the Center, orienting the research objectives, and forming the daily practices of the researchers. There were no takers for such changes, especially given all the work required to put into effect the proposed biological disciplinary interfacing. When we pressed the point, we were often ignored by the senior members or met with overt hostility from younger scientists who saw our interventions as an encroachment on their time and career goals. We explore and conceptualize these dynamics in part 2.

The NSF's first-year site visit (2007) required a great deal of preparation. This preparation proved to be excessive given that the time allotted to Thrust 4 was roughly twenty minutes over two days. The bulk of the materials presented by each of the thrusts essentially recapitulated the materials in the original grant proposal. Since Thrust 4 was not included in that proposal, we had more work to do.

Few suggestions were returned to us, none of which strengthened our position within the Center. Although officials at the NSF had enthusiastically agreed that new forms of post-ELSI collaboration were needed, they had little idea of how to review and evaluate such forms. And although design and invention of experimental practices in engineering and biology were mandated and expected, the privilege of experimentation was not extended to Human Practices. Given all this, it was disappointing, though not surprising, that familiar deliverables were demanded (i.e., policy recommendations derived through the "application" of "principles"), as were additional justifications of our work according to familiar instrumental criteria.

By early spring 2007, six months into our participation in the Center, it became clear to us that the labor of justifying our position within the Center, our research program, and our vision for collaborative ethics constituted an ever-receding horizon.

2 Principles of Design 2006–2007

From Bioethics to Human Practices

The various genome-sequencing projects of the 1990s were significant in providing a first approximation of the core molecular information about the genome. They were no less significant for the ways in which they contributed to a reconfigured moral imagination and thereby to altering relations among and between biology, ethics, and anthropology. From the outset, the genome projects and the bioethics programs affiliated with them traded on the notion that the genome contained the determinative essence of human identity. The run-up to the announcement of the mapping and eventual sequencing of the human genome was replete with the rhetoric of revelation: in reading our DNA, genomic scientists were uncovering the "blueprint" of life, the "holy grail" of biology. Of course, from the outset such rhetoric provoked contestation and rebuttal, but even then it was taken seriously and its proponents succeeded in setting the points of debate and communication.[1] As such, a good deal of anthropological and ethical energy was spent working to imagine, understand, and critically evaluate the supposed capacities and threats introduced by massive genomic-sequencing projects.

Ethics: Technology and Equipment

In a major innovation, federal funds—the largest ethics project in human history, as one actor put it—were devoted to the design and implementation of legal and cultural methods, procedures, and practices adequate to the challenges posed by the sequencing projects.[2] Like the molecular technologies under consideration, these legal and cultural interventions were designed to achieve specified ends.[3] We call this mode of intervention and its standardization "equipment." Equipment connects a set of *truth claims*, *affects*, and *ethical orientations* into a set of practices. These practices, which have taken different forms historically, are productive responses to changing conditions brought about by specific problems, events, and general re-configurations.

The first National Commission, for example, was established in part in response to the abuse of research subjects of medical research. The commission was mandated to develop practices by which research subjects could be protected. The form these practices took was guided by the following considerations: a *truth claim* (human beings are subjects whose autonomy must be respected), an *affect* (outrage at the abuse of such infamous research projects as the Tuskegee experiments), and an *ethical orientation* (human subjects must be protected from such abuse in the future through the guarantee of their free and informed consent). With genomics, more than the autonomy of subjects appeared to be at stake. For many, human nature as well as the integrity of nature more generally seemed threatened. Thus, the sequencing projects contributed to a growing sense that bioethics urgently needed new means, designed to protect human beings from violations of their nature. Neither the affect of concern nor the desire to restrict genetic interventions was new, to be sure. What was new, however, was a growing sense that bioethics as it functioned in authorized spaces, such as government commissions, needed to be recalibrated to meet these new conditions. The means of that recalibration are what we are calling "equipment."

Whereas the protection of research subjects involved the development of regulations upstream from research in the form of Institutional Review Boards and protocols for obtaining informed consent, human genomics appeared to require the design of a set of downstream practices. The objective of this equipment, this pragmatic mode of intervention and regulation, was to mitigate "social consequences" by restricting those directions and applications of research thought to pose a threat to the dignity of human beings. In the United States, equipment of this kind began to be elaborated as part of the Human Genome Initiative ELSI (Ethical, Legal, and Social Implications) project; it has been most thoroughly conceptualized and developed by the current President's Council on Bioethics. The architect and first chair of that council, Professor Leon Kass, proposed a truth claim, an affect, and an ethical orientation for the construction of such equipment: (1) Bioethics matters precisely because what is at stake in biotechnology is humanization or dehumanization, that is to say, the essence of human being is on the line; (2) this state of affairs should inspire a measure of fear and vigilance, for in the face of scientific advance, the "truly" human might be sacrificed; and (3) given the risk of dehumanization, the task of the ethicist is to discover what is truly valuable about human life in advance of any particular scientific endeavor and secure it against scientific excess.[4]

At the beginning of the twenty-first century, after two decades of genomics, it is now clear that the significance of biology for the formation of human life is more than molecular; today we are faced with new forms of the long-standing challenge of understanding living organisms and their milieus. New developments in the biosciences must be accompanied by the invention of new ethical and anthropological analysis and equipment. Focusing on the new Synthetic Biology Engineering Research Center with which we were associated, SynBERC, we argue that contemporary developments call for new forms of collaboration among ethics, anthropology, and biology. Collaboration is a form of engagement appropriate to the shared stakes of biological research and the broad assemblages within which such research is situated. It is animated by the recognition that ethicists, anthropologists, and biologists are working in a shared field of problems. Collaboration therefore requires more than observation and advice, more than submission to oversight. Collaboration requires a reflection on and adjustment of basic work habits.

Post-Genomics: Human Practices

Under the leadership of Leon Kass, the President's Council was oriented by the view that bioethics must begin its work by identifying the "defining and worthy features of human life" so as to determine whether or not those features are put at risk by innovations of biomedical technology. Several characteristics of this orientation are noteworthy. First, these features of human life are universal and ahistorical, that is, they obtain regardless of context or situation. Second, this means that they can be identified without reference to scientific developments. Third, as such, the defining features of human life serve as criteria by which particular scientific programs can be judged as threatening or not to "truly human" life.[5]

When the design of equipment starts with the supposition that science can only pose threats to the integrity of human nature, it is difficult for ethical understandings of *anthropos* to take into account the knowledge produced by contemporary molecular biology or anthropology. Ethics thereby would be positioned exterior to both biological and human sciences. Such positioning makes it more difficult to incorporate scientific knowledge in formulating the stakes and significance of contemporary human practices. Rather than excluding continuing scientific insight from our understanding of the human, it seems imperative to engage molecular biology and other sciences in order to learn what they can tell us about living beings. If one accepts this dialogic and contingent form of engagement, then scientific developments themselves prompt the question: Are contemporary forms of ethical equipment required today? And what critical stance—in the sense of assessing legitimate limits and forms—is appropriate toward and within such contemporary equipment?

Molecular biology demonstrated that DNA is shared by all forms of life and is a remarkably pliable molecule. This means, on the one hand, that if there are questions to be posed about the qualitative distinctiveness of living beings—and there are—such questions must be posed at a different level than the molecular. On the other hand, it suggests that DNA can be manipulated without violating any laws of nature or deep ethical principles per se. Longer and longer DNA sequences are being constructed ever more efficiently and economically each year. Sequences are being inserted with increasingly precision and forethought into organisms; knowledge and know-how are accumulating about ways to make these organisms function predictably. What is at issue for the science, the ethics, and the anthropology is not the metaphysical purity of nature but the biological function of DNA sequences, the extent to which these sequences can be successfully redesigned, and ways in which these redesigns contribute to—or are nefarious to—well-being understood as a biological, anthropological, and ethical question.

Living beings are complex in part because of their evolutionary history; they survived or perished under specific selective pressure in particular environments. Although the products of natural selection demonstrate fitness, this does not mean that this is the only way that the organism can function. Quite the contrary, while evolution certainly contains lessons about organic functionality, for contemporary biologists there is nothing sacred about the evolutionary paths followed to arrive at the functionality. Furthermore, functions themselves are neither inviolable nor immutable. For biologists, there is no ontological or theological reason per se why specific functions—whatever their history—cannot be redesigned. Biologists indeed are making new things. And while this may not violate any sacrosanct ontology of nature, it does not mean that anything goes. It is precisely because we do not think that nature is by essence immutable that these practices and the objects they produce must be carefully examined. The effects of redesign do contribute to a problematization of things (ontology) that must be taken up, thought about, and engaged (ethics and anthropology).

By the end of the 1990s, not only were genomes being sequenced with regularity and a steady flow of genes inventoried and annotated, but an array of other active biological parts and functions is being identified and cataloged. All of this science and technology proceed on the basis of a tacit faith in a principle of an economy of nature. That is to say that nature must consist in isolatable and discrete components and functions. Many biological functions appear to be irreducibly complex in part because the capacities to analyze them, to break them down into parts, do not yet exist. One strategy to address this impasse is to invent the skills necessary to reconstruct those parts and make them function. It is that path—of analysis and synthesis— that is currently being grouped under the rubric "synthetic biology" and that concerns us here.

Today, in the early years of post-genomic science, the insufficiency of what has been called the "gene-myth" is now clearer. It has not been as frequently recognized, however, that the sufficiency of the standard bio-ethical models that arose alongside the discourse of molecularization must itself be exposed to renewed questioning and reformulation. Questioning and reformulation does not mean jettisoning; much of existing bioethical equipment continues to serve a necessary function. It is simply prudent and consistent with our principles for those of us inventing new ethical forms based in *phronêsis*, practical wisdom, to learn from the strengths and limits of previous practices. Limiting the intersection among ethics, science, and anthropology, however, to either upstream bureaucratic review or down-stream impact regulation now appears poorly adjusted to the current situation of dynamic contingency and critical exploration in the biological and human sciences. In sum, loyalty to past practices can inhibit an ability to identify and analyze new challenges. We must take seriously the ways in which current transformations in scientific research modulate past problems as well as the equipment that had been invented to handle them.

Ethical equipment like that developed by the President's Council remains in an ambivalent relation to bioscientific innovations. Strikingly absent from the development of this equipment is any attempt to incorporate the insights of contemporary science into definitions of what it means to be human. We hold that bioethics, as currently practiced in official settings, tends to undervalue the extent to which ethics and science can play a mutually formative role. More significant, it undervalues the extent to which science and ethics can collaboratively contribute to and constitute a good life in a democratic society.

We have been convinced throughout the human practices experiment that specifying which norms are actually in play and how they function might contribute positively to emergent scientific formations. It is worth seeing if such participant-observation can be effectively realized by conducting ethical inquiry in direct and ongoing collaboration with scientists, policy makers, and other stakeholders. Our experiment turned on seeing if within such collaborative structures a better practice could be invented.

Synthetic Biology: The Engineering Ideal

As we signaled in the introduction, synthetic biology began as a visionary but minimally defined project. Its visionary scope and scale, however, need to be underlined so that their audacity and utopian aspirations, and the eventual transition to actual research programs, remain visible. Thus, in 2006, synthetic biology's visionaries could state the following without irony or hesitation:

> Synthetic Biology is focused on the intentional design of artificial biological systems, rather than on the understanding of natural biology. It builds on our current understanding while simplifying some of the complex interactions characteristic of natural biology.
>
> Those working to (i) design and build biological parts, devices and integrated biological systems, (ii) develop technologies that enable such work, and (iii) place the scientific and engineering research within its current and future social context.[6]

Synthetic biology's chief architects approached synthetic biology as a process of modularization and standardization; it appears to us to be developing in and renovating a tradition nicely labeled the "Engineering Ideal in American Culture."[7] Synthetic biology aims at nothing less than the (eventual) regulation of living organisms in a precise and standardized fashion according to instrumental norms. There is a feeling of palpable excitement that biological engineering has the capacity to make better living things, although what that would mean beyond efficiency and specification opens up new horizons of inquiry and deliberation.

Today the engineering project of building parts that either embody or produce specific biological functions and inserting them in living organisms is at the stage of moving from proposal to concept. The concept is being synergistically linked to an ever-expanding set of technologies and to increasingly sophisticated experimental systems. There is agreement within the synthetic biology community that a necessary if not sufficient initial step required to further this project is to conceive of, experiment with, organize, and reach broad consensus on standardized measures and processes. The very qualities of living systems that make them interesting to engineering—that they are robust, complex, and malleable—also make them extremely difficult to work with. The extent to which these difficulties can be productively managed remains to be seen. In any case, at present hoped-for standards are recognized to be initially crude and will certainly have to be reworked in an ongoing manner, but the important step is to begin to create them and to instill an awareness and sensitivity among practitioners as to their importance.

Synthetic biology arose once genome mapping became standard, new abilities to synthesize DNA expanded, and it became plausible to direct the functioning of cells. Its initial projects address a part of the global crisis in public health—malaria. At the same time, the first ethical concerns that it has had to deal with arise from the risk of bioterrorism (see below). The synthetic biology community is obliged to bring these heterogeneous elements into a common configuration. Put schematically, synthetic biology can be understood as arising from, and as a response to, new capacities, new demands, and new difficulties that oblige, in an urgent manner, contemporary ways of thinking and experimenting with vitality, health, and the functioning of living systems. Those investing in the development of synthetic biology expect that it will play a formative role in medicine, security, economics, and energy, and thereby contribute to human flourishing. Questions about what constitutes flourishing and the extent to which synthetic biology can indeed contribute to it are basic and, more importantly, remain unanswered.

SynBERC: A Prehistory

Several core synthetic biology projects were well under way prior to the organization of SynBERC. Two of these were particularly important for the development of Thrusts 1–3. The first is a project at Berkeley, led by SynBERC director Jay Keasling. The project's goal is to take a molecule, artemisinin, that is found in the bark of a Chinese tree and is one of a small group of molecules that remain effective against malaria, and to engineer a system in which the molecule can be produced at a cost that is many times less than the extraction from the tree. This basic work has been accomplished—it is grown in yeast or *E. coli* through a reengineering of the pathways of these common single-celled organisms. So, synthetic biology, at least in this form, exists and it works. The major criticisms of the project come from those who have the legitimate concern that too much hope is being invested in a combination therapy based on a synthetic version of artemisinin that is likely to lead to its potential overuse and the consequent acceleration of resistance to it, with tragic results. That is a valid public health argument, and those holding this position do not advocate eliminating this source, only thinking about consequences.

The partner chosen to take Keasling's work out of the lab and into those regions of the world where it is most urgently needed is another distinctive NGO, OneWorldHealth. The concept around which this NGO is organized is that hundreds of millions of dollars have been spent in research and development in the pharmaceutical and biotechnology industries that have yielded scientific insight, technical improvements, but often no commercially viable product. Their strategy has been to acquire (at the lowest possible cost) the intellectual property generated by this investment and work and to transfer it to countries like India where it can be adapted to local circumstances. The goal is to make available therapeutic advances that might be effective but are deemed to be not profitable enough for multinational pharmaceutical companies. The quid pro quo is for those receiving the intellectual property not to compete in the same markets.

Although it is hard to imagine how one could argue that one should not encourage the development of new anti-malaria drugs in a world in which over a million people die each year from the parasite simply because the molecule to be used in therapy would be produced by reengineering pathways in yeast or *E. coli*, this does not mean that no critical questioning should go on. But critical questioning requires knowledge and understanding. Hence, it is valid to argue that an over-abrupt use of a mono-therapy in a situation where the pathogen is highly adaptive is not a prudent strategy. And the synthetic biologists accept that criticism and are seeking to build the molecule so that poly-therapies that will reduce the likelihood of swift resistance can be built into the design (artificial, organic, natural, and emergent). Surely, changing the genome of yeast to produce artemisinin seems prudent and urgent, knowing full well that it is being designed to be introduced into the bodies of human beings and will thereby change both their internal milieu, which already consists of multiple genomes (both contemporary and archaic) as well as the external milieu in which they live.

So, perhaps unique attention to the question of existing cultural understandings of nature and science at times can obscure other potentially more significant problems and questions. For example, what is perhaps most distinctive about this project is its funding and institutional setting. There is government research money, there is venture capital funding, there is university support, and the artemisinin project is funded in large part by the Bill and Melinda Gates Foundation. This foundation—with a massive gift from financier Warren Buffet—has the largest endowment of any philanthropy in the world. It, like a few other new foundations—Google now has a for-profit foundation—are seeking to assemble health, science, policy, accountability, profit, delivery systems, management styles, scope, and timing in a distinctive fashion. Here is a very American assemblage with global reach. Its norms of productivity and accountability differ from those of the WHO or other such organizations in which national and international politics play such a distinctive part. This assemblage would certainly seem to be making a difference. And that diagnosis implies that we are obliged to think about its significance.

A second important project that was under way prior to SynBERC is located at MIT. It is devoted to building—or learning how to build or to find out to what it extent it is possible to build—standardized biological parts, devices, and platforms. Its goal is to have a directory of such functional units available for order online (http://partsregistry.org/Main_Page) and to make them available worldwide on the basis of an open-source license developed by a nonprofit called Creative Commons. The core concept and initial work has taken place at MIT under the leadership of professors Drew Endy and Tom Knight, integral members of the SynBERC initiative. One original organizational contribution, led by Randy Rettberg, has been to organize an international student competition, iGEM (genetically engineered machines), that has grown exponentially since its inception to include the participation of well over a hundred teams.[8]

Whereas the Keasling project raises one set of ethical and policy problems, the work at MIT poses a different order of challenge. Recent innovations in synthesis technology vastly expand the capacity to produce ever-larger specified sequences of DNA more rapidly, at lower cost, and with greater accuracy. These innovations raise the stakes of the so-called dual-use problem (the idea that technologies can be used both constructively and destructively), expanding existing fields of danger and risk. The relation between technical innovation and the expansion of danger has long been identified in the world of genetic engineering. Previously, these trends have been framed as issues of safety, which can be addressed through technical solutions. To date a number of reports focusing on the governance of synthetic biology have adopted this framing.[9]

It has become clear, however, that not all challenges associated with synthetic biology can be dealt with through technical safeguards. For instance, changes associated with contemporary political environments, particularly new potential malicious users and uses (i.e., terrorists/terrorism), and increased access to know-how through the Internet exceed technical questions of safety. Such challenges cannot be adequately addressed using existing models of nation-specific regulation. New political milieus produce qualitatively new problems that require qualitatively new solutions. In addition, we must confront the challenge of uncertainty characteristic of all scientific research. Although some risks are presently understood, we lack frameworks for confronting a range of new risks that fall outside of previous categories. Such frameworks would need to be characterized by vigilant observation, forward thinking, and adaptation.

Given these conditions, synthetic biology calls for a richer and more sustained inquiry and reflection than is possible in a study commission model of collaboration, wherein formal interaction ceases with publication of a report. To date, work in bioethics has largely consisted either of intensive, short-term meetings aimed at producing guidelines or regulations, or standing committees whose purpose is limited to protocol review or rule enforcement. By contrast, we are committed to an approach that fosters ongoing collaboration among disciplines and perspectives from the outset. The principal goal of SynBERC's Human Practices thrust is to design such collaboration. This enterprise aims at giving form to real-time reflection on the significance of research developments as they unfold and the environments within which research is unfolding. The aim of such collaborative reflection would be to identify challenges and opportunities in real time and to redirect scientific, political, ethical, and economic practices in ways that would, hopefully, mitigate future problems and contribute to human flourishing.

Human Practices 2006–2007

As indicated in the introduction, certain sections of this book are presented as they were originally written. We leave them in the present tense, so as to preserve the state of play at those conjunctures. The following section is a prime example of this mixing of temporalities and is indicative of our understanding of the possibilities and limits of our human practices undertaking as our experiment progressed.

Principles of Design

Within collaborative structures, practice can be oriented and reoriented as it unfolds. This work is accomplished not through the prescription of moral codes, but through mutual reflection on the practices and relationships at work in scientific engagement and how these practices and relationships allow for the realization of specified ends. Straightforwardly: Ethics and anthropology can be designed so as to help us pause, inquire into what is going on, and evaluate projects and strategies. The goal of the Human Practices thrust is to design, develop, and sustain this mode of collaboration. Given that goal, our wager is that the primary challenge for the Human Practices thrust is the invention of diverse forms of equipment requisite for the task. If the scientific aims of synthetic biology can be summarized as the effort to make living things better and to make better living things, then the principal question that orients our efforts to invent contemporary ethical equipment is this: *How should complex assemblages bringing together a broad range of diverse actors be ordered so as to make it more rather than less likely that flourishing will be enhanced?*

We do not yet know what form contemporary equipment will take. At this early stage of our work, however, three fundamental design principles appear worthy of elaboration and testing: *emergence, flourishing*, and *re-mediation*. In initial experimentation these design principles appear to be both pertinent and robust. They are pertinent in that they form part of the research strategies of the biologists and characterize the assemblage of relations within which the research is developing. Initial indications have shown them to be robust in closely related domains (e.g., biosecurity). In these domains they have made visible unanticipated problems and inter-connections, thereby opening up new and more appropriate modes of in-tervention and reflection. One of our initial aims is to test the robustness of these principles in synthetic biology.

Principle 1: Emergence

Research in human practices is underdetermined. Past bioethical practices often operated as though the most significant challenges and problems could be known in advance of the scientific work with which these chal-lenges and problems were to be associated. Our hypothesis is that such practices are not sufficient for characterizing the contemporary assem-blage within which synthetic biology is embedded. This assemblage is a contemporary one: it is composed of both old and new elements and their interactions. While some of these elements are familiar, the specific form of the assemblage itself, and the effects of this form, can only be known as it emerges. We understand *emergence* to refer to a state in which multiple elements combine to produce an assemblage whose significance cannot be reduced to prior elements and relations. As such, the problems and their solutions associated with synthetic biology cannot be identified and ad-dressed until they unfold. Questions concerning what it means to make life different, what it means to make living beings better, and what metrics and practices are appropriate to these tasks can best be addressed in real time as challenges arise and breakdowns happen. The knowledge needed to move toward the desired near future will be developed in a space of relative un-certainty and contingency. Adopting a vigilant disposition that is attentive to a mode of emergence is at the core of our work. In sum, our equipment must be designed such that it generates knowledge appropriate to states of emergence.

Principle 2: Flourishing

In the 1990s bioethical equipment was designed to protect human dignity, understood as a primordial and vulnerable quality.[10] Hence its protocols and principles were limited to establishing and enforcing moral bright lines indicating which areas of scientific research were forbidden. A different orientation, one that follows within a long tradition but seeks to transform it, takes ethics to be principally concerned with the care of others, the world, things, and ourselves. Such care is pursued through practices, relationships, and experiences that contribute to and constitute a *flourishing* existence. Understood most broadly, flourishing ranges over physical and spiritual well-being, courage, dignity, friendship, and justice, although the meaning of each of these terms must be reworked and rethought according to contemporary conditions.

As emphasized in the introduction, the question of what constitutes a flourishing existence, and the place of science in that form of life, how it contributes to or disrupts it, must be constantly posed and re-posed in such a form that its realization becomes more rather than less likely. In sum, the design challenge must be oriented to cultivating forms of care of others, the world, things, and ourselves in such a way that flourishing becomes the mode and the *telos* of both scientific and ethical practice.

Principle 3: Remediation

The third design challenge is to develop equipment that operates in a mode of *remediation*. The term *remediation* has two relevant facets. First, it means to remedy, to make something better. Second, remediation entails a change of medium. Together, these two facets provide the specification of a specific mode of equipment. When synthetic biology is confronted by difficulties (conceptual breakdowns, unfamiliarity, technical blockages, and the like), ethical practice must be able to render these difficulties in the form of coherent problems that can be reflected on and attended to. That is to say, ethical practice remediates difficulties such that a range of possible solutions become available. In sum, our challenge is to design contemporary equipment that will operate in a mode of remediation. This equipment must be calibrated to knowledge of that which is emergent and enable practices of care that lead to flourishing.

We do not presume to know in advance of its actual scientific work how synthetic biology will inform human life. We are persuaded, however, that ethical observation and anthropological analysis are capable of contributing positively to the overall formation of synthetic biology. We think that our contribution can only be effectively realized if this work is conducted in direct collaboration with scientists, policy makers, and other stakeholders. Standard approaches have sought to anticipate how new scientific developments will impact "society" and "nature," positioning themselves external to and downstream of the scientific work per se. The value of collaboration is that it constitutes a synergistic and recursive structure within which significant challenges, problems, and achievements are more likely to be clearly formulated, successfully evaluated, and changed. Following our design principles, our goal is to invent new sets of contemporary equipment, put them into practice, and to remediate things as they unfold.

Human Practices Manifesto 2007

In light of the governance structures under which we were obliged to carry out our experiment, we redoubled our efforts to sharpen and foreground the scientific and ethical stakes of our enterprise. This meant, above all, clarifying the fact that the purpose of our experiment with synthetic biology, and thus of our design for human practices, was neither to contribute directly to the improvement of the biologists' and engineers' scientific work, per se, nor directly to the hoped-for prosperity that might one day issue from the undertakings of which SynBERC is a part. Rather, the purpose was to pose and re-pose the question of how synthetic biology is contributing or failing to contribute to the ethical near future, and to take up the challenge of determining how, and under what conditions, the human sciences might be able provide an answer to such a question.

From the fall of 2006 to the spring of 2007, our designs for Human Practices began to take shape. These designs followed from our diagnosis of the more general post-ELSI situation of bioethics (as described above and as expanded in the next chapter in a more critical and conceptual manner) as well as the specific conditions of synthetic biology as an emergent field (begun here and expanded in analytic detail in chapter 4).

Through this diagnostic work, we honed in on the particularities of SynBERC as the venue in which we would be working. In light of the design principles and the first stages of our experiment (and experiences), we produced a diagnostic manifesto as a means to the practical and conceptual work that lay ahead.

....

Our work is oriented to the goals, practices, and experiences of the synthetic biology community broadly and SynBERC in particular. We are addressing the question: How is it that one does or does not flourish as a researcher, as a citizen, and as a human being? Flourishing here involves more than success in achieving projects; it extends to the kind of human being one is personally, vocationally, and communally. As a placeholder, we note here that *flourishing* is a translation of a classical term (*eudaimonia*) and, as such, a range of other possible words could be used: thriving, the good life, happiness, fulfillment, felicity, abundance, and the like.* Above all, *eudaimonia* should not be confused with technical optimization, as we hold that our capacities are not already known and that we do not understand flourishing to be uncontrolled growth, progressivism, or the undirected maximization of existing capacities. Adequate pedagogy of a bioscientist in the twenty-first century entails active engagement with those adjacent to biological work: ethicists, anthropologists, political scientists, administrators, foundation and government funders, students, and so on. Contemporary scientists, whether their initial dispositions incline them in this direction or not, actually have no other option but to be engaged with multiple other practitioners. The only question is how best to engage, not whether one will engage. Pedagogy teaches that flourishing is a lifelong formative process, one that is collaborative, making space for the active contribution of all participants.

Our goal is to design new practices that bring the biosciences and the human sciences into a mutually collaborative and enriching relationship, a relationship designed to facilitate a remediation of the currently existing relations between knowledge and care in terms of mutual flourishing. The means to inquire and explore to what extent these new relationships will be fruitful consist in the invention, design, and practice of what we refer to as *equipment*. *Equipment* is a technical term referring to a practice situated between the traditional terms of *method* and *technology*.

If successful, such equipment should facilitate our work in synthetic biology (understood as a human practices undertaking) through improved pedagogy, focused work on shared problem-spaces, and the vigilant assessment of events:

Pedagogy: Pedagogy involves reflective processes by which one becomes capable of flourishing. Pedagogy is not equivalent to training, which involves reproduction of knowledge and technique. Rather, it involves the development of a disposition to learn how one's practices and experiences form or deform one's existence and how the sciences, understood in the broadest terms, enrich or impoverish those dispositions.

Events: A second set of concerns involves events that produce significant change in objects, relations, purposes, and modes of evaluation and action. By definition, these events cannot be adequately characterized until they happen. Past events that have catalyzed new relationships between science and ethics include scandals in experimentation with human subjects and the invention of

---

*See ARC Studio 04, "Fieldwork in Affect: Diagnosis, Inquiry, Reconstruction," at http://anthropos-lab.net/studio/episode/04/ (accessed May 3, 2011).

equipment to limit them, the promise of recombinant DNA and its regulation, crises around global epidemics and significant biotechnological interventions, the Human Genome Sequencing Initiative and the growth of bioethics as a profession, and 9/11 and the rise of a security state within whose strictures science must now function. Just as scientists are trained to be alert to what is significant in scientific results, our work is to develop techniques of discernment and analysis that alert the community to emergent problems and opportunities as they take shape.

Problem-space: Events proper to research, as well as adjacent events, combine to produce significant changes in the parameters of scientific work. These combinations of heterogeneous elements are historically specific and contingent. At the same time, they produce genuine and often pressing demands that must be dealt with, including ethical and anthropological demands. In sum, our understanding of the contemporary challenge is to meet what Max Weber calls "the demands of the day" through the design and development of equipment. Such equipment must be adequate to remediating these heterogeneous combinations, the problems raised, and a near future in which it would be possible to flourish.

Our initial task is to provide a set of conceptual tools adequate for an analysis of this problem-space so as to reflect in a rigorous fashion on its ethical significance and ontological status, as well as to provide equipment that contributes to solutions that are more responsive and responsible.

Our experiment concerns the relations among and between knowledge, thought, and care, as well as the different forms and venues within which these relations might be brought together and assembled. Our commitment is anthropological, a combination of disciplined conceptual work and empirical inquiry. Our challenge is to produce knowledge in such a way that the work involved enhances us ethically, politically, and ontologically. Such a project obliges us to think. Thinking no doubt involves the work of "freeing up" possibilities: demonstrating contingency precisely where necessity is expected. But in zones where contingency has become dominant, where heterogeneous truth claims abound, where stable relations have become unstable, and elements, old and new, are being reassembled, thinking must shift modes. Thinking in such a case involves the work of contributing to the form of the near future, scientifically and ethically. Such form-giving, we are persuaded, should be oriented, guided, and evaluated by the hope and goal, the metric, of mutual flourishing. That today we have barely any idea of what such flourishing might consist in only underscores the urgency and joy of undertaking the challenge.

It follows that our challenge is to invent and to practice new forms of inquiry, writing, and ethics for anthropology and her sister sciences and to invite others to do likewise. The dominant knowledge production practices, institutions, and venues for providing an understanding of things human in the twenty-first century are derisory when measured against the ethical, political, and ontological significance of such work. Thinking requires sustained work on the self, with all this requires in terms of adjustments in modes of reasoning and the venues whose mandates are to foster thought. The human sciences are at an ethical impasse: how to connect knowledge of things human to care of things human. To use the classical term, the human sciences are in need of *paraskeue*—equipment. To restate the challenge: Anthropology and her sister sciences are in need of new forms of inquiry and equipment. In that light, we have taken up the experimental work of imagining, designing, and putting into practice one mode of remediating the conditions of contemporary human scientific knowledge production, dissemination, and critique. Such a mode is currently being put "to the test of reality, of contemporary reality, both to grasp the points where change is possible and desirable, and to determine the precise form this change should take."* It is an experimental mode: an anthropology of the contemporary.

Progressing in this direction entails changing the metrics and forms of current practices, habits, and affects. Above all, it entails recursive experimentation and learning of a collaborative sort. In its initial stages, *experimentation* simply means trying out different configurations of inquiry, critique, and colabor, and then evaluating those practices and their results in a manner such that one can learn from these experiences. *Recursive* means punctual assessment and reconfiguration of those efforts. *Collaboration* means inventing new forms of work that redistribute individual and collective contributions and limitations. Redistribution alone, of course, is insufficient: such work must be *remediative*; it must remedy significant dimensions of current pathologies through diagnostic analysis of the current state of things, followed by the design and practice of pathways operating in a different mode and in a modified medium. Such pathways are designed so as not to inflame the wounds of *ressentiment* that plague the academy through more *ressentiment*. Rather, they are designed to realize the hope and goal of mutual enrichment, of flourishing, as we have already suggested. Here we are merely insisting that the question of what constitutes a good life today, and the contribution of the life sciences and the human sciences to that form of life, must be vigilantly posed and re-posed.

*Michel Foucault, "What Is Enlightenment?" in *Ethics: Subjectivity and Truth*, ed. Paul Rabinow (New York: New Press, 1997).

3 Interfacing the Human and Biosciences 2007

Three Modes

As of 2007, we identified four strategic tendencies that self-identify as synthetic biology. There are two whose goal is to engineer whole cells. One seeks to do this from the "top down" by simplifying existing organisms and then engineering whole genomes or chromosomes, inserting them into the existing cellular machinery so as to orient them to function in a specified manner.[1] Another "bottom-up" approach, following in the line of earlier efforts at creating synthetic life-forms, attempts to build proto-cells from the amino acids up.[2] The two other variants are the ones we are working most closely with. The distinction between these two variants is an analytic one that we draw from our observations. It is not a stated or otherwise formalized distinction. The first has been developed primarily by a group of researchers at MIT. It consists of the attempt to engineer, modularize, and standardize working parts on the analogy of industry and prior developments in engineering. The goal of the MIT researchers is to make synthetic biology an engineering discipline in the formal sense.[3] The second is a variant of this approach, one that works in conjunction with the MIT model. It is characteristic of researchers at UC Berkeley, where there is a stated openness to the goal of standardizing synthetic biology as an engineering discipline, but where actual work focuses more on specific functional problems. This approach seeks to develop and use synthetically engineered parts not as a goal in themselves or as the demonstration of the power of the subdiscipline. Rather, techniques and the work of standardization are taken up insofar as they can be made to contribute to work on specified bioengineering projects. Such projects are not, strictly speaking, limited by the label of synthetic biology.[4]

DIAGNOSIS: THREE MODES OF ENGAGEMENT

Our own research and reflection, our reading of the relevant literature in sciences studies, and insightful inquiry from intellectual historians have convinced us that there is—and has been—a level of pragmatic concern and development that lies between technology and method. Settled technologies honed to maximize means-ends relationships abound in our industrial civilization; the social and biological sciences have produced vast reservoirs of methodological reflection to justify and advance their work. Yet inquiry into past situations of change as well as contemporary explorations makes clear that neither technology per se or grander methodological elaborations quite cover the terrain of how diverse domains are brought together into a common assemblage. Nor do they sufficiently explain how ethical considerations and demands have been brought into a working relationship with the quest for truth and made to function pragmatically—that is, they do not account for equipment.[5]

In what follows, we provide a diagnosis of three current equipmental modes. We are developing a diagnosis that is designed to be directly helpful for our work at SynBERC but is also intended to be applicable (with appropriate adjustments) to a range of analogous problem-spaces. This approach has helped us to analyze more clearly the challenges of how to proceed in organizing and putting into motion this multidisciplinary endeavor. The diagnosis offered in this chapter discriminates the ways in which various *modes of engagement* are designed to manage and respond to qualitatively different kinds of problems. In this way, distinctive modes of engagement can be interfaced and adjusted to each other such that the resulting assemblage is adequate to the kinds of problems that SynBERC and other similar contemporary enterprises are designed to address.

This chapter analyzes two predominant modes of engagement, the *representation of technical experts* and the *facilitation of "science and society,"* as well as a third mode, emerging today, which we call *problematization and inquiry*. A goal of this analysis is to explore the conditions under which existing expertise and "boundary organizations" can be appropriately adjusted and interfaced with synthetic biology, with each other, and with the third mode, which is emergent and in the process of design and experimentation.

For each of the modes, we present a list of "externalities" and of "critical limitations" as a series of talking points. There is a substantial scholarly and professional literature on many of these issues. Rather than giving the impression that we are comprehensively presenting each of these questions, we prefer a schematic form as a means of indicating that these are topics we are attempting to think about, explore, and draw lessons from at this initial stage of both our inquiry and the development of SynBERC. At the end of the chapter, we raise a series of challenges that those attempting to work collaboratively must face.

To aid the diagnosis and design of possible interfaces among these modes, we begin with an ideal typical and schematic presentation of these modes so as to determine practices that are helpful and unhelpful in order to determine more clearly existing limitations and challenges. Our approach is in the line of the construction of "ideal-types" proposed by Max Weber a century ago. We build three distinct forms that are constructed so as to be analytically distinct one from the others. We are fully aware that in the "real world" these divisions are not so neat and compartmentalized. The function of the ideal-type, after all, is to highlight distinctions so as to enable inquiry into the specifics of existing cases. At the same time, of course, these ideal-types have been constructed from materials drawn from preexisting efforts and examples. Hence there can appear to be a slippage between the ideal typical function of producing an analysis and a description of existing configurations.

Mode 1: Representing Modern Experts

Mode	Key Externality	Critical Limitation
Representing Experts	Emergent Problems	Metric of Uncertainty

Mode 1 consists in inventorying, consulting, and cooperating with experts. The core assumption—often taken for granted and not subject to scrutiny— is that the expertise of existing specialists in one domain is adequate without major adjustment to emerging problems. Of course, in many instances, an adequate pairing of problem domain and expertise does exist. The vast number of technical specialists trained and supported by the state bureaucracies, corporations, international agencies, and nongovernmental nonprofits of the industrialized world certainly are competent to address many current challenges. It is worth remembering that many of these challenges have been formulated, worked over, and compartmentalized by these experts and the organizational form, practices, and limits within which they operate.

Expert knowledge functions as means-ends maximization. Even when such expert knowledge is operative, it gains its very strength precisely from its capacity to bracket purposes or goals. Expert knowledge is structured and functional only when that which counts as a problem is given in advance, stabilized, and not subject to further questioning. In emergent situations, however, neither goals nor problems are settled, and so technical expertise cannot be effectively marshaled without some adjustment. In many instances, obviously, when goals and problems become settled, technical expertise must be given a useful place within an assemblage. Said another way, routinization is normal but qualitatively different from states of emergence or innovation.

Having access to technical competence and successfully deploying it in delimited situations (which need to be identified and stabilized themselves) so as to effectively address problems are not the same thing. Hence, in addition to technicians, in stable organizations there is a need for managers or technocrats whose task is to oversee and coordinate specialists and technicians. Such coordination facilitates a *cooperative* mode of engagement by subdividing specializations and assigning tasks. As with technical expertise, it is frequently supposed that the competencies of technocrats are transferable from stable to emergent situations. As such, in the United States technocrats and technicians often rotate out of public, governmental, or corporate service, and take up positions as consultants or lobbyists claiming transferable competence.

In Mode 1, the role of the social scientific practitioner is to identify and coordinate legitimated specialists and technocrats. The Mode 1 practitioner is expected to maintain broad overview knowledge of a number of subdisciplines at least to the extent that the Mode 1 practitioner can legitimately claim to present a range of candidates as authoritative and available. Candidates are presented and ranked along scales (both formal and informal) of authority, availability, connections, and character. The Mode 1 practitioner's authority is based on this work of inventory and ranking. The types of equipmental platform according to which Mode 1 practitioners calibrate their work are those that distinguish among kinds of authorized experts and draw these experts into a cooperative frame.

The Mode 1 practitioner frequently does not provide (or take as part of the job) a critical analysis of the status of expertise per se, or of existing expertise and its specific functions. Rather, the Mode 1 practitioner understands his work as providing an evaluative assessment of specific first-order practitioners, in a first-order mode. The metric of this inventory-making is not a second-order one. We are taking the distinction between first- and second-order observers from the German sociologist Niklas Luhmann. A second-order observer is someone who observes observers observing. This sounds opaque but is actually quite straightforward. First-order observers take their world as it comes to them (often in a highly mediated form). They then do their work. This intervention in the world is what Luhmann refers to as "observing," hence the term is more than perceptual; it is an action, frequently a sophisticated one. A second-order observer observes actors acting. Such a second-order action is neither removed from the world nor given any special privilege. Furthermore, as Luhmann writes, "A second-order observer is always also a first-order observer inasmuch as he has to pick out another observer as his object in order to see through him (however critically) the world."[6] We take up this distinction in a nonjudgmental and simple manner: it helps to distinguish different positions and different modes of doing one's work.

The Mode 1 practitioner *represents* existing *expertise*. This representation takes a twin form. The Mode 1 practitioner literally re-presents existing expertise in a readily comprehensible form (often PowerPoint). The Mode 1 practitioner is a representative for the legitimacy of existing expertise. The Mode 1 practitioner does not put forth claims of validity concerning substantive issues dealt with by the chosen experts. It follows that under specific circumstances, in fundamentally stabilized situations, institutions, and problems, Mode 1 work can provide benefits by identifying, bringing together, and representing existing expertise.

From the outset of the formation of SynBERC, it was clear that even in the domains where the existing core of specialist expertise might well be more directly pertinent (e.g., intellectual property) than in some others (e.g., ethics), it was certain that start-up companies with whom scientists at SynBERC had direct association as founders or board members (e.g., Codon Devices, Amyris) would have ready access to such experts (i.e., would have already taken great care to address intellectual property issues). This supposition has been amply supported by the evidence. In a word, the small start-up companies associated with SynBERC and other parallel organizations have already hired patent lawyers and have given priority to related financial matters (or in the case of established organizations such as BP have whole departments long in place). These counselors are privy not just to the generalities of synthetic biology as an emergent field but to the specifics of the scientific and technological inventions at issue. Further, venture capitalists who have invested in these start-ups provide the contacts necessary for maximizing protection and insist on their enforcement.

Finally, SynBERC was conceived within a certain ethos of maximizing the "commons" and was associated in a working relationship from the inception with groups such as Creative Commons with long experience in innovative patent and organizational design. It is generally recognized that questions concerning industrial strategies and IP are of fundamental importance to synthetic biology. Work to date has focused on how synthetic biology will have to adapt its open-source goals to existing models of industrial strategies and IP. Our approach is to inquire into what distinctive forms of industrial partnerships and IP can be invented given the objectives of *specific synthetic biology projects*.

Externalities: The Price to Be Paid

There are some immediately identifiable externalities that bear on Mode 1. The term *externality* as we are using it is taken from neoclassical economics. It refers to factors that "result from the way something is produced but is not taken into account in establishing the market prices."[7] The identification of such externalities allows one to pose the question: When and where is it an effective use of limited resources to undertake a Mode 1 practitioner's strategies?

1. In emergent problem-spaces, appropriate experts do not necessarily exist. This fact falls outside of Mode 1 operational capacities. Such a deficit, however, does not imply that there is no possible way to adjust and integrate existing expertise. Rather, it simply calls for second-order reflection on this state of affairs.
2. Even when appropriate expert knowledge does exist, its very strength—technical criticism as means-ends maximization—gains its legitimacy precisely from its capacity to bracket purposes or goals. In an emergent situation, such bracketing must itself be subject to scrutiny.
3. In either case, a different skill set is required to move into the contested networks and pathways of what is taken to be the impact, consequences, opinions of "society" or "the public." The response of Mode 1 practitioners to this challenge is to look for other specialists in surveying opinion, assessing consequences, and preparing for the impact. The reservations of #1 and #2 thus apply here as well.
4. Mode 1 is based on the modernist assumption that there is a society, that it has been divided into value spheres, that there is a problem of legitimation, and that the challenge is to invent a form of governance in which these issues can be adjudicated through procedure and specialization. These assumptions have been debated and challenged for over a century.[8] And within a new globalized, accelerative, security *oecumene*, it is not obvious, far from it, that Mode 1 presuppositions are defensible.

Critical Limitations: Structural Incapacities

When unacknowledged, externalities can become *critical limitations*—that is, they can introduce structural incapacities. Given Mode 1 externalities, the question needs to be addressed: Where expertise is engaged, what are its critical limitations? By answering this question, we will be able to pose the question of where and when Mode 1 experts are useful in an assemblage such as SynBERC. We have identified several critical limitations:

1. In Mode 1, the future appears as a set of possibilities about which decisions are demanded.[9] The range of these decisions is delimited by a zone of uncertainty. The genesis and rationality of such a zone is that Mode 1 experts operate with a metric of certainty. The ever-receding zone of uncertainty, however, is not fundamentally unknowable, only uncertain. But precisely because it forms a horizon depending on current decisions, this zone of uncertainty cannot be specified in advance. Uncertainty, however, does not undermine the decision-making imperative of experts. Rather, it compels incessant decisions and affirms that an appropriate form of verifiable certainty (probability series, risk analysis, technical measurements, etc.) can be attained. The authority of experts is not undermined by the oft-demonstrated inability either to forecast the future or to make it happen as envisioned. Rather, this dynamic provides the motor of their legitimation. In sum, a zone of uncertainty is an intrinsic part of this equipmental mode.[10]
2. In Mode 1, uncertainty is taken up as a boundary condition. It allows Mode 1 practitioners to move from the generation of verified claims and their delimitation to the coordination of discussion and communication. Rather than deflating the authority of experts or making obvious the need for other modes of inquiry, this move to discussion and communication allows for the rehearsal of the past triumphs of expertise, and renders such past verificational successes as points of reference to orient debate about the present and near future.[11]
3. Uncertainty entails an ever-receding horizon. As such, rather than functioning as a fundamental limitation, uncertainty provides a refinement and corrective such that Mode 1 practitioners can (ostensibly) operate more realistically and, therefore, more effectively. Mode 1 practitioners attempt to factor in "uncertainty" as a parameter in identifying and ranking expertise. What they fail to factor in is the structural insufficiencies of existing expertise both external and internal.

Mode 2: Facilitating Relations between Science and Society

Mode	Key Externality	Critical Limitation
Facilitating Science and Society	Selecting Stakeholders	Formal Proceduralism

We take the distinction between Mode 1 and Mode 2 from the work of Helga Nowotny and coauthors, who have been part of an active debate and an articulated conceptualization of the strengths and limitations of Mode 1.[12] Their book is an elaboration of a report commissioned by the European Commission. In fact, Mode 2 has become the norm for official policy in Europe in regards to "science and society." Although there are examples of this mode in the United States (see below), such instances are dispersed and are not currently normative in an official policy sense. Mode 2 arose as a reaction to the perceived arrogance of scientists and technocrats and their lack of professional competence to deal with concerns beyond their direct disciplinary or subdisciplinary questions. Further, policy makers and civil society activists have discovered and documented the inability of Mode 1 to include a range of existing social values in planning; the honest admission that neither the purely scientific nor the technological per se were competent to evaluate consequences and impacts; that by including opinion both as a set of numbers produced by polling and surveying techniques, projects could be better designed so as to meet less resistance and be more representative.

Mode 2 social science practitioners are *facilitators*. Their role *qua* facilitator is to bring heterogeneous actors (scientists, technical experts, policy makers, lawmakers, civil society actors, political activists, industry representatives, government and private funders, etc.) together into a common venue. That venue, often created for a particular crisis or event but eventually standardized and routinized, is fundamentally a space for *representation* and *expression*. Stakeholders are encouraged to express themselves, to advocate, to denounce, to articulate, to clarify, and eventually, it is hoped, to form a consensus. Such a consensus is taken to be normative and made to function equipmentally in the organization of research and development programs.

Mode 2 is calibrated according to an innovative equipmental platform. This platform takes "social values" as norms for discriminating which activities are appropriate, and elaborates these values such that they can serve as the basis for the organization of such activities. It follows that the challenge for Mode 2 practitioners is to develop procedures for identifying significant social stakeholders, discerning their opinions and values, and designing mechanisms through which such opinions and values become normative for research and development in the sciences. In order to meet this challenge, venues that function to facilitate boundary organization and its modes of governance must be invented and institutionalized. Both in Europe and in the United States, the venue for this work has predominantly been the "Center."[13]

Cutting-Edge Example: Nanotechnology and Society

A leading example of Mode 2 social science is the Center for Nanotechnology in Society at Arizona State University (CNS-ASU). CNS-ASU has been designed to take on board and adjust its organizational practices to the limitations of Mode 1 by focusing on "emerging problems" and "anticipatory governance":

> Designed as a boundary organization at the interface of science and society, CNS-ASU provides an operational model for a new way to organize research through improved reflexiveness and social learning which can signal emerging problems, enable anticipatory governance, and, through improved contextual awareness, guide trajectories of NSE [nanotechnology science and engineering] knowledge and innovation toward socially desirable outcomes, and away from undesirable ones.[14]

The proposed means of moving toward such socially desirable outcomes is to "catalyze interactions" among a representative variety of publics. The metric of these interactions is not to produce technical expertise per se, but to raise the consciousness and responsive capacities of high-level policy makers, scientists, and "consumers."[15] Interactions and awareness are facilitated by designed and monitored dialogues on the goals and implications of nanotechnology. This engagement will facilitate the construction of a communications network positioned upstream rather than downstream of the research and development process. Upstream positioning is designed to anticipate and evaluate the impact of nanotechnology on society before "rather than after [its] products enter society and the marketplace."[16]

Equipmental Platform: RTTA and Reflexive Governance

Two sets of strategies are being designed and developed in order to meet CNS-ASU's goals. The first is a program of "research and engagement" called "real-time technology assessment" (RTTA). RTTA consists of four components:

1. "Mapping the research dynamics of the NSE enterprise and its anticipated societal outcomes."
2. "Monitoring the changing values of the public and of researchers regarding NSE."
3. "Engaging researchers and various publics in deliberative and participatory forums."

4. "Reflexively assessing the impact of the information and experiences generated by our activities on the values held and choices made by the NSE researchers in our network."[17]

The second procedure is a program for "anticipatory governance." Anticipatory governance can be distinguished from "mere governance," defined as "the kind that is always found running behind knowledge-based innovations." Rather, through the facilitation of interfaces between societal stakeholders and researchers, CNS-ASU is attempting to develop practices of governance with the capacity to

1. "Understand beforehand the political and operational strengths and weaknesses of such tools."
2. "Imagine socio-technical futures that might inspire their use."[18]

Externalities: The Price to Be Paid

There are some immediately identifiable externalities that bear on Mode 2. The identification of such externalities allows one to answer the question: Where and when should Mode 2 strategies be undertaken in synthetic biology?

1. Mode 2 attempts to factor in and move beyond the limitations of Mode 1's focus on existing expertise. However, given built-in funding and legitimacy demands, such a move is frequently hindered. For example, the identification and management of polling such diversity itself requires further experts. Yet additional specialists are required to manage these burgeoning classifications, groups, and subgroups. Audit culture expands to meet its own criterion of inclusiveness, accountability, and responsibility (bureaucratic demands of accountability): the technologies of polling and opinion collection to be developed and managed by experts (in polling, in the presentation of results, in public relations, etc.).[19] In sum, the challenge of moving beyond expertise encounters the requirement for new experts.
2. Mode 2 supposes that ethical science is science that benefits society, which is made up of stakeholders, whose values must be given a venue for expression. Such supposition generates two problematic limitations. The first is that various stakeholders are vulnerable to the charge of being ignorant or not competent: scientists often believe that laypeople are incapable of understanding the details of their work in its own terms (often correct) and hence are not capable of producing

legitimate evaluations (often contestable). Policy makers, social activists, and social scientists often believe that the results presented in scientific or technology journals do not correspond to the complexity of social reality. Journalists' attempts to explain science to society are thought to simplify both poles. It follows that charges and countercharges of hype joust with charges of ignorance.

3. The second limitation is that it has become clear in Europe that techniques of producing society's representatives were required, as well as techniques of legitimating these representatives. The legitimating process is frequently challenged by those who consider themselves to be excluded.

Critical Limitations: Structural Incapacities

Given these externalities, the question needs to be addressed: Once the appropriate venues for Mode 2 have been established, what are its critical limitations? By answering this question, we will be able to pose the question of where and when Mode 2 practitioners are useful in a collaboratively normed assemblage such as SynBERC.

Mode 2 is characterized by at least three identifiable critical limitations:

1. Mode 2 takes seriously the challenge to respond to society and the public in order to orient research responsibly. However, experience has shown that specifying who exactly one is talking about when one references *society* and/or the *public* frequently turns out to be an elusive task.[20] These broad rubrics cover highly diverse actors.

2. Furthermore, two decades of work in STS and related fields have put into question the very existence of referents to such homogenizing terms like *science*, *society*, and *public*. Sciences are plural when they retain any distinctiveness at all. *Society* has been increasingly replaced by *community* and the *individual* in its neoliberal frame.

3. In Europe, given the bureaucratic framework of the European Union, not surprisingly, the way in which the first critical limitation has been dealt with has been through the channels of representation and formal procedures. Proceduralist approaches, however, rarely resolve value disputes, although they may provide means of adjudication in specific instances. Likewise, proceduralist approaches rarely resolve scientific differences. Finally, proceduralist approaches tend to mask power differentials.

4. Regardless of how successfully bureaucratic procedures are designed and implemented, problems remain. As many critics have pointed out,

such as the President's Council on Bioethics in the United States and Attac in France, opinion polling, formal proceduralism, consensus building, and the multiplication of representatives' expression cannot answer the ethical and political questions of whether or not a given course of action is good or bad, right or wrong, just or unjust. In fact, proceduralist exercises have no way of posing these questions. It follows that representation and expression as modes of organizing scientific and political practice, much like technical expertise, while coherent and valuable within a democratic framework, nonetheless because of their inherent limitations possess serious dangers that must be taken into account.

Mode 3: Problematization and Inquiry

Mode	Key Externality	Critical Limitation
Problematization	Cooperative Engagement	First-Order Deliverables

Mode 3 has the challenge of how best to identify the limits of each mode and to design interfaces among and between them. When an enterprise is confronted by difficulties (conceptual breakdowns, unfamiliarity, technical blockages, and the like), ethical practice should be able to render these difficulties in the form of coherent problems that can be reflected on and eventually attended to. That is to say, Mode 3 takes up difficulties such that a range of possible solutions become conceivable. Mode 3 is oriented to the near future. It begins in situations rife with blockages and opportunities: the challenge is to conceptualize significant problems emergent in these situations. Such conceptualization is a step toward giving form to the near future as a series of problems in relation to which possible solutions become conceivable.

Mode 3 work does not privilege claims to being "revolutionary" or even "cutting edge" as what is distinctive and intriguing about developments in synthetic biology and other emergent domains. Such tropes are modernist ones drawn from a prior historical configuration that focuses attention on what is "new" and "radically transformative" as the locus of significance. Mode 3 work, in contradistinction, is drawn to the combination and recombination of elements old and new into an assemblage whose parameters do not turn on its newness per se. Rather, in what has been described elsewhere as "the contemporary" (as opposed to "the modern") what counts as significant are the forms and possibilities that open up once the quest for the new is moderated and backgrounded (although not ignored).[21] Which objects are taken up and how they are combined and recombined into an assemblage is part organizational, part conceptual, part technical—and part pragmatic. Posing the question of how such an assemblage might be best put together, made to function effectively, designed to cope with breakdown and unexpected occurrences, and to discern emergent problems is a principal objective of Mode 3.

Mode 3 works to provide a diagnostic of the near future. Its challenge is to remain situated among contemporary blockages and opportunities so as to reformulate these blockages and opportunities as problems. Mode 3 work proceeds by identifying the ways in which formerly stable figures and their elements are becoming recombined and reconfigured. In sum, problematization taken up as a Mode 3 task is oriented toward the remediation of current blockages and opportunities by conceptualizing the near future as a series of problems in relationship to which possible solutions become available to thought.

Human Practices Parameters: Adjacency, Collaboration, Reconstruction

In the early stages of Human Practices, the design and development of Mode 3 consisted in facing three primary challenges.

The first Mode 3 challenge was to specify its positional parameters as *adjacent* and *second-order*. *Adjacent* is defined as "situated near or close to something, especially without touching." Our task at SynBERC has been to remain situated close to the work of both the biological engineers and the policy-oriented human scientists while carrying out Mode 3 work. We undertook to establish such an adjacent position by remaining resolutely committed to what Niklas Luhmann has called a "second-order" observational position. By "observation," Luhmann means more than a passive gaze. Observation, rather, equally entails active intervention in a situated manner as an essential component to the production of sociological insight. Hence, second-order does not mean removed, but rather demands a design for participant-observation: the active observation of the practices of those with whom one is engaged.

This adjacent and second-order stance takes the practitioners and products of Modes 1 and 2 as primary objects of observation. Such critical attention facilitates identifying strengths and limitations of Modes 1 and 2. Given the positionality of Modes 1 and 2 as consultative and first-order, it is not surprising that even sympathetic practitioners of these modes would put forth the demand for first-order and advisory deliverables from Human Practices. Consequently, a primary challenge for Mode 3 is to develop a tool kit of responses and practices that temper and reformulate such demands.

To the extent that such demands can be tempered and reformulated, it becomes possible for second-order problems to be collaboratively worked on. Such collaboration certainly will require existing experts. However, the expertise will need to be interfaced with emergent problems in such a way that experts will be required to think forward rather than reproduce existing insights. In sum, our work is oriented toward understanding how potentially viable design strategies emerge, how these strategies might inform synthetic biology, and what efforts are undertaken to integrate them into a comprehensive approach to the near future.

A second Mode 3 challenge thus concerns formulating parameters for *collaboration*. Given the emergent character of innovations and practices in synthetic biology, the precise forms of collaboration have not and cannot be settled in advance. Rather, such collaborations will require intensive and ongoing reflection with SynBERC PIs on emergent ethical, ontological, and governance problem-spaces within which our work is situated and develops. We have been experimenting with both directed group meetings as well as having undergraduate and graduate students directly engaged within SynBERC labs as their work unfolds.

The aim and purpose of such collaborative engagement is *reconstruction*. In Human Practices, we follow John Dewey's specific technical meaning:

> Reconstruction can be nothing less than the work of developing, of forming, of producing (in the literal sense of that word) the intellectual instrumentalities which will progressively direct inquiry into the deeply and inclusively human—that is to say, moral—facts of the present scene and situation.[22]

What is pertinent in Dewey's formulation is that science and ethics are interfaced and assembled in accordance with the demands of "progressively directed inquiry." Such inquiry is not primarily directed at real or imagined consequences or first-order deliverables, although the work of Modes 1 and 2 on these topics is both relevant in and of itself as well as primary data for reflection. Rather, inquiry is directed at the possibility of the invention and implementation of equipment that facilitates forms of work and life.

Mode 3's challenge is daunting. Older patterns of practice with their power inequalities as well as dispositions formed under different conditions continue to remain in place. Hence, whether such collaborative and reconstructive facilitation could occur and whether they might prove to be efficacious or beneficial remain to be seen.

Externalities: The Price to Be Paid

There are a number of immediately identifiable externalities that bear on Mode 3. The establishment of clarity about external and internal limits distinguishes warrantable scientific advance from opinion and hype. The identification of such limits allows one to pose the question: When and where is it an effective use of limited resources to undertake Mode 3 strategies?

1. Mode 3 is allied with, but should be carefully distinguished from, the Foucauldian analytic practice of the "History of the Present."[23] When analysis is undertaken with that goal, its task is to show the lines flowing back from the present into previous assemblages (and elements and lines that preceded those assemblages). Such work functions to make clear the contingency of current expert knowledge, its objects, standards, institutions, and purposes. The goal is not primarily to debunk or delegitimate such expertise, although a dominant mode of academic criticism habitually does takes the form of denunciation. Rather, the goal is to make clear how such expertise came about, what problem-space it arose within, what type of questions it was designed to answer, how and where it had been successfully deployed, and what blind spots were produced by its very successes. The purpose of analytic work in the History of the Present is not necessarily to replace the specialists and managers that already exist. Above all, it aims to open up current practices to critical scrutiny.
2. The habits of elite scientists as well as the institutions and ethos of bioethics orient expectations toward a mode of cooperation, not collaboration.

Critical Limitations: Structural Incapacities

Given these externalities, the question needs to be posed: What are the critical limitations of Mode 3? The range of critical limitations of Human Practices is not yet known. However, two limitations can be identified at the outset:

1. Mode 1 and Mode 2 are designed to work within and be facilitated by governmental, academic, and other stabilized venues. These are legitimate venues when the equipmental demands consist in the regulation or regularization of a problem-space. Well-characterized equipment exists for operating in such non-emergent spaces. Adaptations to

emergent domains of work such as synthetic biology and nanotechnology are under way.

2. There will be a repeated and insistent demand for Mode 3 practitioners to provide expert opinion, propose first-order solutions, represent opinions, invent and implement a venue for expression, and facilitate consensus. Mode 3 practitioners acknowledge the validity of such demands for certain problem-spaces, actors, and venues. Mode 3, however, is designed such that fundamentally it cannot—and should not— honor such requests insofar as it operates on and in emergent problem-spaces. There clearly is a price to be paid for respecting this externality. It is the price to be paid for being patient, consistent, and clear-sighted. This consistency may well add value eventually to Mode 3. Its immediate worth, however, is found in its bringing attention to the need for inquiry.

Conclusion: Interfaces

What if, as seems likely given the premises of the strategy of designing and constructing appropriate forms, there actually were no experts in emergent domains and problem-spaces? Thus, for example, while everyone would have agreed readily that at the time of SynBERC's founding there were no specialists in the first three thrusts—Parts, Devices, and Chassis—there were nonetheless scientists with diverse skills needed to establish such innovative and coordinated specialization. Developing venues, modes of practice, technical and other equipment, modes of collaboration, and so on, are, after all, central goals of the Center. The founding strategy was to identify a challenge, make its significance comprehensible, and pursue strategies for addressing it. There was an excited confidence that with success, others— many others—would follow. A new mode of practice would be launched.

Human Practices: Interface with Mode 1

Logically, it follows that, as with Thrusts 1–3, so, too, with Thrust 4. Simply cooperating with technical experts and keeping a watchful eye on the engineers and biologists seemed and seems to be an insufficient, even an implausible, way to proceed. Indeed, such an approach seemed and seems likely to provide the false assurance of short-term deliverables and the potential for strategic misdirection over the longer term. Consequently, an obvious initial challenge has been to invent venues within which academic experts at a distance, who might otherwise only share a cooperative relation to emergent hybrid emergent assemblages such as SynBERC, are situated in such a way that their existing expertise can be remediated and redeployed in view of new problems. The claim is not that existing experts have nothing to offer. The question to be explored is: What can Human Practices provide that existing experts themselves cannot?

One of the distinctive organizational characteristics of SynBERC was its division into thrusts; another was its strategy to include "test-beds" from the start. A test-bed is a concrete research project designed to function as a proof of concept for work in the thrusts. Originally there were two of these—bacterial foundries, tumor-seeking bacteria—and then a third, biofuels. The Berkeley and MIT Thrust 4 leaders agreed that, informally, the MIT group (and its Mode 1 collaborative approach) would serve, in addition to its other contributions, as a test-bed for the Berkeley group's experiment in inventing a new type of equipment. With this division of labor, it was hoped that a collaborative approach could be developed. The advantage of this strategy was that the Mode 1 team would produce immediately recognizable deliverables: workshops, conferences, specific recommendations, organizational advice, network connections in the power centers of the East Coast, and so on.

It was clear that initially Mode 3 would have no such list of familiar deliverables or modes of delivering them. What we did have, however, was a keen sense (based on years of anthropological research in the world of biotech and genomics, contemporary reflections on that world, and deep experience in ethical work in the broader political and industrial context) that current modes of practice had built-in structural limits and, because of the very way they had emerged and been institutionalized, were unlikely to be flexible and creative enough to collaborate effectively within an organization such as SynBERC. We took as an initial task a rigorous diagnosis of what such change might look like, and the initial steps toward actualizing such change. Of course, no one knew in advance if the scientific test-bed form would produce successful collaboration with the separate thrusts.

Human Practices: Interface with Mode 2

If a primary task of Mode 2 is to facilitate representation and expression of stakeholders, this work is likely to be relevant at a subsequent stage. As fields such as synthetic biology have barely begun to take shape, to gain funding and attain a visibility arising from their accomplishments as opposed to the positive or negative hype that surrounds such enterprises, it is likely to be the case that the "public" or "society" may well have no opinion whatsoever, and certainly no detailed opinion or well-informed representatives (none exist) at the early stages of emergent disciplines and assemblages.

There are now professionals at organizing public opinion and alerting stakeholders in other assemblages of how they might or should be concerned about developments in related fields. These analogy professionals' claims to be representing broad numbers of people and civil society interests should be examined with care. That being said, these Mode 2 professionals have already established funding mechanisms, relationships with journalists, functioning websites, networks with heterogeneous civil society groups, and so on. It would seem to be a pressing and legitimate function of Mode 2 practitioners to assess, sort, adjudicate, and moderate emerging common places and rhetorical thematics.

If preemptive analogizing is both rampant and low on the serious speech act metric, equally futurology is not the answer to emergent things. There are many versions of predicting or narrating the future. Among them is forecasting. Forecasting refers to the use of quantitative analysis to identify the future trajectories of current trends. The goal of such forecasting is to anticipate small variations from these trends (e.g., variations in oil prices). Forecasting has two built-in limitations. First, it bases its conclusions on the logical outcomes of only one possible future. Second, this one possible future is thought to be a direct and predictable unfolding of current states; as such it assumes a much greater similarity between the present and the future than usually proves to be the case. Forecasting as a way of dealing with the future requires assembling technical experts who can quantitatively elaborate extensions of current trends. If the future is contingent and emergent as in zones such as synthetic biology, however, such forecasting has limited value.

4 Synthetic Biology 2008

From Manifestos to Ramifying Research Programs

During 2007–8, efforts at SynBERC (and in other allied venues) shifted from writing and disseminating manifestos and grant proposals to facing the challenges of animating research programs. Synthetic biology was *ramifying*. To *ramify* means to produce differentiated trajectories from previous determinations. This stochastic process produces unexpected effects that "may complicate a situation or make the desired result more difficult to achieve."[1] As synthetic biology shifts from manifestos to research programs, its initial directions, distinctiveness, and results can begin to be specified and characterized. Our task is to document and analyze these ramifications as they unfold. Our purpose is to make these trajectories available for critical understanding and evaluation. And, we hope, for eventual remediation. Our initial task has been to specify and characterize the ramifications of various strategies for moving from manifestos to research programs both in the biosciences and human sciences. In chapter 6, our analysis of research programs in the human sciences will be expanded.

SECTION I: IN SEARCH OF SYNTHETIC BIOLOGY— FOUR RESEARCH PROGRAMS

The great danger of analogy is that a similarity is taken as an identity.[2]

From the inception of synthetic biology through 2007, researchers at MIT provided the framing vision that has been adopted by funders and used as a device for scientific journalists, and that orients the way in which many (but not all) researchers discuss synthetic biology. This framing confidently envisions synthetic biology as a nascent engineering discipline predicated on its self-proclaimed ability to produce standardized and interchangeable parts. However, upon closer examination, not surprisingly what is actually taking place in these research centers and their labs is more complicated.

Four Research Programs: Parts, Pathways, Genomes, Systems

	Problem	Analogy	Venue	Human Practices	Externality/ Critical Limitation
Parts	Standardizing Biological Units	Computer Systems	iGEM	Regulated Commons	Non-enforceable Venue
Pathways	Designing Synthetic Pathways	Microbial Chemical Factories	Agile Assemblage	Cooperative Specialists	Nonrecursive Pathways
Genomes	Designed Genomic Platforms	Cloning	Lab Fab (building prototypes)	Safety-by-Design	Technological Reductionism
Systems	Regularizing Biological Cybernetics	Heuristic Use	Traditional	Moral Contract	Insufficient Attention to Collaboration

Most of the players in synthetic biology agree on the need for (a) rationalized design and construction of new biological parts, devices, and systems as well as (b) the redesign of natural biological systems for specified purposes, and that (c) the versatility of designed biological systems makes them ideally suited to solve challenges in renewable energy, the production of inexpensive drugs, and environmental remediation, as well as providing a catalyst for further growth of biotechnology.

However, what is understood by these goals is quite diverse. In fact, those assorted understandings are currently contributing to different ramifications of synthetic biology. In order to clarify this situation, we distinguish at least four design and composition strategies currently operating under the name of synthetic biology. In the section that follows, we provide a characterization of each of these four strategies, first considering their biological research programs. Then, we provide a synopsis of efforts within these programs that fall under the human practices rubric. We pay special attention to externalities (those costs expected to be paid by someone else) as well as critical limitations (the—often unacknowledged—range of structural capacities and incapacities) of each research strategy. We should emphasize for the reader that we are once again presenting these strategic orientations as ideal-types.[3]

1. Parts

The first and most widely publicized type has been formulated by researchers at MIT and is exemplified by the BioBricks Foundation (http://biobricks

Problem	Analogy	Venue	Human Practices	Externality
Standardizing Biological Units	Computer Systems	iGEM	Regulated Commons	Non-enforceable Venue

.org/). This approach has two goals. The first goal is to transform biology into a fully standardized and abstracted engineering discipline understood in a literal sense on the *analogy* of electrical and computer engineering. The second goal, in line with the first, is to reduce biological systems to modular and additive parts, which can be combined in a linear fashion to form more complex functional units.[4] Such standardized biological parts are the principal objects of interest and investment. The success of this approach depends on the ability to black-box the evolutionary contingency and nonlinear dynamics of underlying biology, just as, or so the analogy runs, the development of computer software succeeded in black-boxing microphysics.[5]

The responsibility for designing the parts-based approach—and publicizing it—has been taken up by engineers (electrical and civil) at MIT, especially Tom Knight and Drew Endy. A unique contribution of their "LEGOs" approach has been the development of the "BioBricks" standard as well as the registry of standardized parts (http://parts.mit.edu). The principal venue for the expansion of this approach is MIT's annual International Genetically Engineered Machines (iGEM) competition (http://parts.mit.edu/igem07/index.php/Main_Page); iGEM brings together a growing international set of undergraduate research teams whose projects, in order to qualify, must meet the BioBricks design standardization criteria and, to qualify for awards, whose parts must be deposited in the BioBricks registry. The iGEM competition constitutes, for the BioBricks approach to design and composition challenges, the central vehicle of expansion and legitimation for establishing itself as the norm for synthetic biology.

The manifestos of the BioBricks approach imagine and plot a comprehensive remaking of the biological sciences. Moving from the scale and scope of their guiding vision to more mundane experimental results, they have encountered research obstacles. Not only are biological processes more difficult to black-box in the lab than in discourse,[6] but the original and innovative venue lacks the power to enforce its standards. It seems clear that an adequately financed fabrication facility, a "parts fab," will be required if the BioBricks approach is to be fully vetted and its range of applicability tested. Such a parts facility would be a stable industrial-scale organization with a clear mandate to produce standardized parts.[7] It would presumably be staffed predominantly with technicians, not undergraduates or postdocs.

Human Practices: Regulated Commons

The BioBricks vision and its manifestos have been the most comprehensive and inclusive of human practices considerations. BioBricks explicitly recognizes the need for innovative rethinking of intellectual property issues, security concerns, organizational form, and ethics. This vision turns on the idea that in order for synthetic biology to be successfully realized, an ethos of openness and collaboration must be fostered from the outset, and venues created for its implementation.[8]

A primary *externality* of the parts approach is that there is no enforcement mechanism connected to the ethos that it proposes as the guiding feature of synthetic biology's vision. In the initial stages, the expansion of practices of openness and sharing has been dependent on the goodwill of participants; insufficient attention has been paid to the pragmatics of organizational enforcement.[9]

One key *critical limitation* of the BioBricks approach is its own tacit resistance to establishing a venue in which human practices participants can play a collaborative and productive role as equal partners. Perhaps a more intransigent obstacle to realizing an ethos of openness is the fact that many of the major players are pursuing other conflicting IP commitments.[10]

2. Pathways

Problem	Analogy	Venue	Human Practices	Externality
Designing Synthetic Pathways	Microbial Chemical Factories	Agile Assemblage	Cooperative Specialists	Nonrecursive Pathways

The first completed project that showed that synthetic biology could be a robust and effective approach is the Keasling Lab's (http://keaslinglab.lbl.gov) design of microbial pathways for the production of an anti-malaria molecule, artemisinin.[11] Although the Keasling Lab is committed publicly to supporting the parts-based approach to synthetic biology, the artemisinin research program was constituted on a different analogical basis. If the analogical basis of the BioBricks approach is computer engineering and the objects it seeks to construct are standard biological parts, the analogical basis of the Keasling Lab's approach is industrial chemistry transferred into the cell (i.e., "microbial chemical factories"), and the core objects—on which it focuses its attention and its resources, and around which it has constructed its facility—are enzymatic pathways.[12]

A distinctive aspect of the Keasling approach is its venue. The artemisinin project, like Keasling's subsequent project on biofuels,[13] is set within an institutional framework that allows research to be directly ramified into practical solutions to real-world problems. The artemisinin project was organized as a collaborative endeavor by specialists from the Keasling Lab at UC Berkeley, the Bill and Melinda Gates Foundation, OneWorldHealth, and Amyris. This approach not only enabled the design and development of new microbial pathways in yeast (and *E. coli*), but required that essential connections be fashioned from the outset among and between strategic partners. As a result, this endeavor set a precedent for the organization of synthetic biology as a collaborator in a multi-institutional approach to addressing some of the most pressing real-world problems. This same approach is now being applied to biofuels at the Keasling-directed Joint Bio-Energy Institute.[14]

A defining characteristic of a pathways-based approach is the study of evolutionary processes so that dynamics such as fitness and variation can be leveraged as part of the design tool kit.[15] Rather than black-boxing biological complexity, evolutionary processes, and variation, this approach embraces them in order to produce specified molecular compounds in an efficient and scalable manner.

If the power of this approach is its problem-driven focus on pathways, this is also its limitation. The production of enzymes and the reconstruction of pathway dynamics are only one set of processes to be learned from evolution. Although this approach has proven successful in producing high-value compounds such as artemisinin, at present it is not formulating a research program that squarely addresses the challenges of constructing yet more complex devices and systems.[16] This remark is not a criticism, only an observation as to the form of the Keasling orientation.

Human Practices: Cooperative Specialists

Currently, the human practices dimension of the pathways approach recognizes the need to engage specialists for managing financial and regulatory matters as well as the work of developing deliverables. The strength of Keasling's venue is that it considers and accounts for this need by building pathways between the lab and other institutions from the outset such that once the scientific milestones have been reached, an apparatus is in place for the translation of the designed pathways into effective solutions.

This arrangement, however, implies an *externality*. It presumes a cooperative division of labor in which its scientific work assumes a linear and unidirectional relation to the rest of the pathway. The other research departments of Keasling's venues have been designed such that developments in any one area of research can rapidly be accounted for and adjusted to in the other areas of research. Unlike these biological research and engineering departments, those specialists tasked with managing human practices issues are downstream and external to the biological research. The price to be paid for such an externality is that while the strength of Keasling's cooperative approach is the anticipation of how to move from the lab to deliverables, its weakness is that if these pathways prove inadequate, there is no available internal mechanism for adjustment. This agile assemblage remains agile only under certain circumstances.

The success of the artemisinin project covers over the fact that these venues are not as flexible and agile as the actors believe them to be. As such, what is taken to be an acceptable externality in one case—that is, a cooperative pathway—is structurally assumed to be sufficient in other cases. A key human practices *critical limitation* of the cooperatively constructed pathways approach is that it is not collaborative. By this we mean that if the original division of problem areas and specialties proves to be insufficiently agile or flexible, there is no internal mechanism to rethink and implement rapid adjustments. This arrangement is likely to prove troublesome in areas where the scientific product, the regulatory challenges, the financing, the mode and ramifications of applications, and their interconnections are not known in advance. For example, artemisinin was identified from the start as the malaria molecule of choice, the appropriate funding was noncommercial, and a nongovernmental agency (experienced in biotech-based health care delivery in developing countries) was available. Had any part of this pathway not been already in place, more human practices input would have been required. By contrast, in an area such as biofuels where none of the components of the proposed pathways are already in place, and where the contours of the field of ramifications are largely unknown, the smoothness of the previous operations is unlikely to be replicated. In sum, a cooperative state of affairs, taken as sufficient for all cases, becomes a critical limitation and not only an externality.

3. Genomes

Problem	Analogy	Venue	Human Practices	Externality
Designing Genetic Platforms	Cloning	Lab Fab (building prototypes)	Safety-by-Design	Technological Reductionism

Another type of research program focuses on the design and construction of "minimal cells." This self-description, however, is somewhat misleading. Actually, the privileged objects of study and intervention in these programs are *synthetic genomes*, which are designed, modified, reconstructed, and synthesized.[17] The analogical basis of these programs is *cloning*. The goal is to fashion synthetic genomes so that they can be inserted into and function within existing cellular hosts. The purpose is to leverage cell functions, including mechanisms' self-reproduction and the capacity for adaptation. This whole-genome approach to synthetic biology is predicated on the assumption that existing cellular machinery will function as a predictable and (ultimately) nonproblematic biological chassis for these designed genomes. A common feature of these approaches is the claim that enough is known (or will be known) about evolutionary processes and genomic biology to proceed with the construction of synthetic genomes designed for specified functions. It is anticipated that genomes would be versatile as a refactoring machine for synthetic biology.

Two examples of labs using this strategy are those of George Church at Harvard Medical School[18] and the J. Craig Venter Institute.[19] Church, a PI at SynBERC, is directing a project to design and construct a minimal genome "capable of replication and evolution, fed only by small molecule nutrients."[20] Given what has been learned from the genome-sequencing projects and from the study of directed evolution, the Church Lab is attempting to build a minimal genome that can function as a safe and controllable chassis.[21] Church's minimal genome offers at least two immediate benefits to synthetic biology. First, it demonstrates a strategy for minimizing the scale of complexity in engineering design. Second, from the outset, it is attentive to issues of safety; it has built-in internal-control mechanisms based on new nucleotides (that don't exist naturally) that the lab has designed specifically for this purpose.

The J. Craig Venter Institute has set as its goal the construction of artificial genomes that serve as multi-flexed platforms capable of receiving (and continuing to function with) a series of specific molecular inserts—genetic "cassettes" carrying designed functions. The goal, one might say, is to build a prototype organic robot.

The Venter Institute has devoted time and resources to charting a wide range of variation and diversity existing in the wild. They have demonstrated that there is an existing dynamic exchange of molecular material in evolutionarily regulated milieus. The documentation of these processes is normative in its use of such milieus to argue that a type of genomic experimentation is a naturally occurring phenomenon going on in the wild with salutary evolutionary consequences.[22] The Venter design and research strategy—as well as its manifestos—is at the opposite pole of BioBricks within synthetic biology's current field of options. Instead of black-boxing biological processes, the Venter Institute approaches evolution as a vast lab

within which a nearly infinite number of experiments past and present provide invaluable lessons of what nature is capable of.[23]

Human Practices: Safety-by-Design

Those currently working on the design of synthetic and artificial genomes
devote attention and resources to issues of safety and security, and what
they take to be attendant social consequences. Their strategy for addressing these human practices concerns can be called "safety-by-design."[24]
There is an explicit effort to design genomes in such a manner so as to have
maximum control over their functionality. Design attention is devoted to
minimizing the risk of survival or reprogrammability outside of the lab.
Safety-by-design's purpose is to fabricate genomes that when circulated,
the effects, both negative and positive, can be accounted for and prepared
for in advance.[25]

The key *externality* of this approach is that it can only address those
aspects of the security challenge that are amenable to technological safeguards.[26] Security issues are framed as problems of dual use in which the
principal challenge arises from the threat of "bad" actors "misusing" technologies created for benevolent purposes. This framing is taken to call for
a technological response by existing specialists: Can a biological chassis be
designed in such a way that it cannot be subsequently "misused"? Other
significant aspects of biosecurity—such as challenges associated with the
current political milieu or preparation for unexpected events, which are not
amenable to safety-by-design—are externalized.

To the extent that this externality is taken to be generally sufficient,
it becomes a *critical limitation*. That is to say, safety-by-design becomes a
critical limitation when it is held that the salient security challenges can
be mitigated adequately through technical means, by police procedures
among and between labs, and trust in the expertise and character of current specialists. Once this externality becomes a critical limitation, there
are no other human practices resources within this venue readily available
for responding to other unexpected and unpredicted ramifications.

Safety-by-design is an attempt to extend self-governance models developed by the 1974 Asilomar Conference and its successors. However,
the success in managing "experiments of concern" depends on the kinds
of venues developed in the 1970s, '80s, and '90s. The scientific, industrial,
and political milieus today are strikingly different. Given the Internet and
the globalization of science, access to materials and specialized knowledge
is widespread. As such, the technical safeguards being developed by those
designing genomes can only have limited efficacy. To the extent that these

technical procedures give the illusion that security issues amount to the management of "experiments of concern," they themselves function misleadingly as "experiments of reassurance," to coin a phrase. We hold that taking such experiments of reassurance as sufficient, explicitly or otherwise, constitutes the significant critical limitation of a safety-by-design approach.

4. Systems

Problem	Analogy	Venue	Human Practices	Externality
Regularizing Biological Cybernetics	Heuristic Use	Traditional	Moral Contact	Insufficient Attention to New Forms

The fourth type of approach in synthetic biology takes as its targeted object neither parts, pathways, nor genomes. Rather, the object of scientific and technological interest is a biological system (often multicellular) understood in an evolutionary milieu. Here the aim of synthetic biology is not only to produce intracellular functions, but includes the goal of intervention and redesign of whole-cell and multicellular systems as well. Its goal is to discover the extent to which abstraction and standardization of bioengineering is feasible at the systems level. New design and composition techniques as well as collaborative strategies are required to pose the question of standardization and abstraction in a manner that will allow them to be approached experimentally.[27]

This approach proceeds by explicitly taking into account the critical limitations of the analogies at work in the other approaches. It acknowledges the heuristic value of analogies from other engineering domains for the provisional orientation and initial design. However, it understands that the use of analogies can be misleading. It follows that, at the level of specifying design parameters, some attention must be paid to the limits of dominant analogies in synthetic biology (computer engineering, microbial chemical factories, cloning) and the extent to which they apply to biological systems.

The Ron Weiss Lab at MIT and the Arkin Lab and the Anderson Lab at UC Berkeley are prime examples of the systems approach in synthetic biology. Many of the so-called "protocell" projects—efforts to design and build minimal cells—are characteristic of this approach as well.[28] A shared strategy across these labs is to test familiar engineering goals such as standardization, decoupling, abstraction, predictability, and reliability for biology. The problem the lab poses is this: Given the seeming complexity and idiosyncrasy of cellular context, the challenge is to account for and abstract from the distinctive characteristics of living systems and to formulate principles of design accordingly.[29] By contrast to the other approaches, the notion of cellular context is made an explicit part of the design strategy from the start, and it is strategically factored into such challenges as the "the functional definition of devices and modules," and the "rational re-design and directed evolution for system optimization." The purpose of such contextual considerations is to make biological engineering modular and predictable at the level of cell populations as well as individual cells.[30] The Weiss, Arkin, and Anderson Labs are distinctive in that they are oriented so as to pose and answer questions about the limits of standardization, while at the same time designing specified research projects that are addressed to real-world problems and applications that contribute to their solution.

Human Practices: Moral Contract

Although there are no explicit statements in the manifestos, personal com-munications and close examination of scientific articles reveal an under-lying ethical substrate in which developments in science and significant medical issues are combined in commitment to the common good. The funding of a series of their research projects reveals a connection and a commitment to medical issues. For example, a project funded by the Cystic Fibrosis Foundation explores signaling systems in bacterial populations so as to design biological interventions that would down-regulate the produc-tion of microbial biofilm, a source of great distress for CF patients.

The Weiss Lab's approach to issues of human practices is in the line of the alliance between patient groups and genome sequencing that was promi-nent in the 1990s. This alliance consists of patient groups providing fund-ing for research projects that, while not being immediately therapeutic or instrumental, hold a plausible promise of identifying and characterizing the underlying biological conditions within which pathologies develop. Moral commitments are addressed in the form of contractual arrange-ments wherein research results are made available to more clinically ori-ented specialists in return for funding.[31]

An *externality* of this moral-contract approach to human practices is that when there are fundamental shifts or blockages, or for that matter successes, there is frequently no built-in capacity for adjustment between the contrac-tual parties. Consequently, the arrangement either dissolves or must be re-negotiated. The researcher in this arrangement is bound by the problems and interests of the patient organization; if his or her own research ramifies in other directions, other sources of funding must be found.

A *critical limitation* of this approach is the tacit assertion that paying the price of externalities frees the research program from having to build collaborative venues within which human practices can function as an integral element of research design and priorities. The moral-contract frame for human practices concerns runs into the critical limitation of presuming that these concerns can be sufficiently accounted for through externalities. There are parallels to this critical limitation in the recent past: the sequencing projects positioned human practices downstream and outside of the design of their own research programs; the parts-based approaches have included human practices at the discursive level but have not involved them in the shift to research programs; the pathways approach has designed and implemented interfaces with human practices specialists, but this is cooperative and nonrecursive; and synthetic genomic design approaches seek to convert all security problems to technical problems as a way of retaining autonomy.

SECTION II: HUMAN PRACTICES — INITIAL RAMIFICATIONS 2008

At the level of manifesto, a distinctive feature of SynBERC has been its inclusion of a component that we called *human practices*. We coined the term so as to avoid the current commonly accepted label, *social consequences*. As we have said, we began with the working hypothesis that an attempt to develop and implement a human practices approach different from ELSI, and the standard policy specialists that it has generated, would encounter conceptual and pragmatic blockages. The reason we anticipated blockages was that we knew that we would be operating in an emergent terrain. And we knew that past expectations, practices, and expertise would be only partially applicable. We also knew that there was a deep conservatism in the policy-oriented branches of the human sciences. Careers change slowly and rewards are generally scant for rethinking categories and practices upon which those careers have been fashioned and sustained. We took our mandate to consist in experimentation and eventual implementation of the collaborative design and development of distinctive approaches to bringing the human and biological sciences into a new set of relationships. Consequently, the shift from manifestos to research programs in human practices required an orientation stage during which we charted the move from manifestos to research programs in terms of organizational ramifications. Developing the diagnostic and analytic tools necessary for this work laid the groundwork for implementation of a research program in human practices as part of a comprehensive program for synthetic biology. Given our working hypothesis, we were cognizant that reflecting on how, where, and when organizations and programs did *not* ramify was another equally interesting, if not always pleasurable, aspect of our inquiry and of our practice.

Reorientation: From Consequences to Ramifications

What is at stake in these shifts from manifestos to research programs? What is the best way to understand them? And how is this shift taking place in human practices? One thing is clear analytically: to characterize the stakes or the process involved as engineering and its *social consequences* is rhetorically misleading and conceptually inadequate. There are several reasons for making this claim. First, none of these programs or centers were funded by the U.S. government in order to engage in the untrammeled pursuit of knowledge. As with the Human Genome Project, they were established to keep sectors of the U.S. economy and its scientific and technological base at the forefront of an ever-more competitive global playing field. That playing field includes concerns with health, environment, and security. All the manifestos and all the grant proposals underscore and justify themselves, at least in part, on this ground. In that sense, what is loosely referred to as society is an unexamined presupposition, not a consequence. Second, all of the NSF-funded centers have a mandate to achieve financial independence after ten years, and all are obliged and pressured to establish relationships with industrial partners as soon as possible. In sum, the idea that a practical discipline like engineering, supported for economic and political reasons, is somehow separate from *social consequences* is misleading.

Of course, many other things will follow from scientific developments: discoveries, blockages, power struggles, patents, career moves, and so on. Some of these will be planned, others not; some predictable, others not; some desirable, others less so. All of this will depend in large part on the degree of success or failure to achieve results, to meet milestones, to raise money, and so on. It is more rigorous to analyze this situation not simply as the cause-and-effect *consequences* of the production of objects and truth claims in engineering disciplines, but as *ramifications* to be analyzed and refashioned. Upon reflection, it is obvious that the very same scientific or technological results could be taken up and mobilized in many different directions. Thus, the *object* of human practices research is ramifications, not consequences; its *method* is observational and analytic; its *mode* is collaborative.

The difference between a *social consequences* approach and a human practices *ramifications* approach can be highlighted by a non-exhaustive list of the factors involved in the creation of Amyris in Emeryville, California.

Although Jay Keasling is one of the founders of the company, Amyris is not a *social consequence* of Keasling's work at Berkeley. Rather, it is a *ramification* of the Keasling Lab's overall research agenda, of the need for increased funding to accelerate promising work in a competitive environment, of the mandated obligation in American universities to commercialize research, of the quest for novel solutions to real-world problems, of the desire to prosper beyond the lifestyle provided by a university salary, of the rewards and incentives of creating jobs, of the taxes generated, of the need to experiment with new hybrid alliances of industry/philanthropy/university partners, of the drive to open new markets, of the pursuit of venture capital funding, of the need to patent products of these hybrid alliances in innovative ways, of the response to growing concerns about the deteriorating state of the environment, health, and security. Each one of these vectors preexisted before any specific research agenda, forms assemblages with the other vectors in distinctive ways, and will partially inflect the fate of the enterprise. This claim does not mean that the production of enzymes through designed divergent evolution is the same as patent contests, only that they both operate within close proximity to each other and are codependent. The patent policy is no more the *social consequence* of the drive to engineer enzymes to break down cellulose than that drive is the *social consequence* of the current structure of the American university. Rather, they form dynamic, sometimes predictable, sometimes turbulent, and, at times, emergent configurations. In sum, the objects that concern us can ramify in different directions depending on events, capacities, obstacles, and interventions. Or the lack of intervention; for, as Niklas Luhmann has taught us, in a situation of contingency—and that is unquestionably the situation of synthetic biology in 2007—non-action is also an action.

Observations: Initial Ramifications

We note that during the first year or so of SynBERC's operations, there have been instances where we have observed the thwarting, tacit or explicit, of potential branching effects, which seemed to be moving in promising directions. An initial task, therefore, was to document and analyze those blockages. We identify three areas of blockage that have been pertinent to synthetic biology's development: (1) intellectual property and open source, (2) security and safety, and (3) moralistic and ethics.

One domain that aroused and sustained interest among the researchers is *intellectual property*, especially patenting. From the outset, the BioBricks Foundation has spearheaded the discussions of creative approaches to open-source strategies. Starting with the analogy of patent issues in software, and drawing directly on the Open Source Initiative templates, BioBricks has held forums on the topic of intellectual property, issued general guidelines and statements of principle, and continues to animate discussion and debate over the more technical standards and norms that are appropriate and productive for the specific concerns of synthetic biology.

We note, however, that the impact of these discourses on actual patenting practices within the larger synthetic biology community and even within SynBERC have been limited. There are a number of reasons for this disjunction between arousal and performance. For example, previous contractual arrangements by the Keasling Lab's artemisinin project with its partners, the Bill and Melinda Gates Foundation and the NGO OneWorldHealth, constrain the partners from making their discoveries (including parts) available to others under an open-source agreement. Other researchers have made arrangements with a growing list of biotech companies that limit their participation in an open-source arrangement. Many of the younger researchers expressed an interest in following this direction. In sum, while a discourse of open sourcing of parts holds a certain normative sway for at least one faction of SynBERC, so far its pragmatic bite has been minimal.

Another key area in which discussion has taken place during the manifesto stage of synthetic biology is *security*. This topic has been given a place of prominence by the National Science Foundation. Although one of the visionaries of synthetic biology, Drew Endy, has worked to put issues of biosecurity on the collective agenda, there has been only a minimal and rather grudging response (again mainly discursive) to his efforts from within the community. Many of the researchers we have talked to, while acknowledging that the topic is considered significant in Washington, D.C., and in the media, do not themselves place security issues high on their own personal agenda. They seem to feel that someone else should take care of it by proposing regulations and safeguards, and they will simply comply with them. Just as bioscientists (and others) have learned to accommodate safety procedures (especially rigorously regulated at the National Labs) as necessary, if minor, annoyances, so, too, our experience has been that members of SynBERC are willing to follow whatever directives might be forthcoming but do not themselves take a proactive stance toward developing such policy. We observe that as with the series of reports on biosecurity sponsored by the Sloan Foundation, the sentiments of concern expressed are sincere, but the actual recommendations are limited in scope, albeit laudable first steps.

Within SynBERC, the most overt attention to *safety* issues has been given by the Church Lab in its work on safety-by-design (the design of chassis that would either be unable to survive in the wild or are designed with multiple fail-safe mechanisms). The strategic orientation of this approach follows in a direct line with that of the Asilomar Conference.[32] The Asilomar strategy can be summarized as follows: Accept the need for public reassurance; establish laboratory safety as the defining characteristic required for such reassurance; strive to achieve autonomy from regulation of an onerous sort through the performance of responsible laboratory procedures and safeguards; equate those safety procedures with security compliance. This strategy has been successful over three decades.

We observe, however, that there is room for doubt as to whether what we call experiments of reassurance—to juxtapose them to the national security focus on "experiments of concern"—actually do provide general safety and security technologies that extend beyond the specific parameters of each individual lab's experimental system. Finally, nothing in this strategy directly addresses in a pragmatic manner the broader issues of the emergent national security environment in the United States and its potential to ramify in directions that synthetic biologists and others would find extremely constraining. The most obvious lacuna is the issue of preparedness: scenario enactments for the aftermath of either an accident or an act of terrorism. The ramifications of such an event would depend on its timing, nature, and effects, but no one can doubt that the ramifications would be significant and enduring.

Initially surprising, but less so upon reflection, has been the tacit acceptance, and even an overt demand by researchers and others within SynBERC, that Human Practices operate in the *ELSI mode*. There is willingness, even an eagerness, among some researchers, to discuss, from time to time, "ethics." By this term, they mean the type of questions and answers, as well as the way in which they were posed and responded to, that were developed in ELSI: Are we violating nature? Are we being responsible? Are we contributing to progress? And today: What would a terrorist think about? The researchers understand these topics in a broadly moralistic frame: Is science good or bad? Were they to take them up as ethical issues, they would have to consider changes in their own practices and habits. This they are reluctant to do; and there are no career or institutional rewards for doing so. Thornier and more intrusive matters such as human subjects requirements and institutional review board (IRB) constraints have not yet been an issue since synthetic biology has not advanced to a stage where it would have to accommodate these regulations and bureaucracies.

This informal, consultative, and communication-oriented mode of interface among researchers and ethicists took shape during the reign of ELSI even if it was not a formal part of its program. The researchers do their work in their university and commercial laboratories (and in consultation with their technology transfer officers, patent lawyers, venture capitalist associates, etc.), and then they convene for conversations at lunch (pizza is often the *plat du jour*) to have a discussion about issues of broad concern and general interest. Not surprisingly, the mood in such a venue is positive: No one is required to attend; no decisions are made; there is no production schedule; there is no accountability; the policy-oriented professors who convene such get-togethers have neither the power nor the authority to make anyone do anything; the discourse of freedom and autonomy of responsible actors discursively produces self-satisfaction—one feels serious and well-rounded discussing matters of concern. As they say in New Jersey: "What's not to like?"

Diagnosis: Incapacity by Design

We argue that these areas of blockage indicate scientific and ethical inadequacy. For the situation and its dynamics to be otherwise, however, power relations and institutional practices would have to be altered in a significant fashion so as to be genuinely open to contestation, collaboration, and rethinking. In order for SynBERC to live up to its manifestos and its mandates, there would have to be a program of research and collaboration that took these considerations seriously and was willing to change practices, habits, dispositions, and career patterns. Today, although it is not unimaginable that such changes could begin to be implemented, it is far from the actual state of affairs. It is no surprise that human practices interventions that would require more difficult thinking and action are less welcomed and less easily accommodated. We identify two reasons why, to date, SynBERC has not been able to integrate its human practices' mandate into its research programs: (1) *trained incapacities* and (2) *current power relations*.

Current practices among the principal investigators of synthetic biology reinforce the ELSI mode of defining and addressing issues as *social consequences*. They position human practices as downstream and external to their research, as capable of contributing in a soft advisory mode or as an extracurricular enrichment-for-the-soul get-together. Surprisingly, this positioning has come not only from the bioscientists, but has been taken for granted by many human scientists. What seems to be operating is not bad faith or ill will. Rather it seems to be an instance of what the great iconoclastic social thinker Thorstein Veblen called "trained incapacity."[33] By this telling phrase, Veblen means simply that those trained in one set of capacities developed for one set of circumstances, and who having been using those capacities over long periods of time, are as a type unlikely to be able to remediate their skill set to accommodate new situations with new problems.

In the case of synthetic biology, such trained incapacity encourages participants to suppose that while the biological challenges of post-genomics require new thinking, new technologies, new forms of organization and interfacing among labs and the like, work in ethics and the human sciences can be conducted using prior modes and prior questions, and can be governed by prior power relations. When trained incapacity is coupled with active disinterest, there is unlikely to be innovation or remediation.

In the recent past, when forceful proposals for structural change that would have constrained or redirected the actions of the PIs within SynBERC have been put forth, they have occasioned vehement disagreement—in and of itself a healthy reaction in a young organization. These disagreements, however, did not result in a rethinking of organizational form or in any refashioning of interfaces. Rather, they resulted in an exercise of sovereign power that substituted itself for the procedures, practices, and ethics of a more representative constitutionalism. Anthropologically, we have observed that sharp exchanges over research issues in the biosciences are a part of the mode in which research as well as business is conducted. When disputes arise from within the Human Practices side of the organization, however, such productive exchange is stymied and even actively resisted. Given the fundamental inequality of the power relations within the organization, intervention by Human Practices has not been well-received. In such instances, we have been told that we do not fully grasp which part of the discourse of collaboration is for official consumption and which part is off-limits to those not within the higher reaches, to use an analogy, of court society with its ranked nobles and its sovereign.[34]

That being said, there were ramifications of an initial confrontation that remain worthy of attention. One significant event that took place in the weeks preceding Synthetic Biology 2.0 (the annual international conference on synthetic biology) is worthy of note. The final day of that conference was to be given to a discussion of security and synthetic biology. A pressing issue confronting the organizers was how to formally constitute the community of synthetic biology researchers so as to institute procedures of self-governance. The immediate reason for attempting this constitutional apparatus was to establish policies for monitoring and regulating purchases of oligonucleotide sequences. Central players in the formation of synthetic biology balked. Extended and vehement behind-the-scenes discussion resulted in the proposal being withdrawn from the agenda. Rather than constituting procedures of self-governance, a proposal was substituted to have the Sloan Foundation fund a study as to how best to regulate the emergent synthesis industry, which all parties agree, with its vastly increased technological capacity to provide longer and longer base pair sequences, constitutes a crucial variable in the wholesale construction and distribution of parts.

The Sloan Report responsibly lays out the issues in detail and provides a valuable baseline for discussion. However, such issues as how to think about transnational governance, presumably one of the most important issues for those concerned with controlling the dual-use potential of synthetic biology, were not seriously addressed. Issues of preparedness were side-stepped as well. In sum, the mode chosen for the governance questions was to outline a (preliminary) technical response, articulated by a handpicked group of experts, vetted by a very small group of actors. This mode of decision making and of community governance is consistent with the drive to autonomy, minimal regulation, and translation of all concerns into technical questions. This mode of decision making and governance has proved to be quite robust, since, it should be noted, it has survived the arrival of the Internet as well as the rise of the biotech industry.

Another result of the sovereigntist intervention was to summarily oust the person who had occupied the "ethics" and "policy" position in the original constitution of SynBERC and to replace him with an anthropologist at Berkeley and a political scientist at MIT. That decision was made with minimal consultation; at the time it seemed to be a plausible choice both because SynBERC itself was bicoastal and because the political scientist was primarily interested in policy issues and the anthropologist in inquiry and critical reflection. Although this choice seemed logical and complementary at the time, the decision has ramified in directions that perhaps could have been better directed had the SynBERC directors taken seriously the difficulty and importance of the task of inventing a form of human practices adequate to synthetic biology.

Conclusion: From Manifestos to Research Programs

Synthetic biology might continue to develop in a mode of trained incapacities, comfortable cooperation, decidedly unequal power relations between the biological and human sciences, and its focus on social consequences, which, by definition, remain external to the biological research per se. There is little question that such a direction is one that the majority of engineers and biologists involved are comfortable with—to the extent that they have any position at all on these issues. There is also little question that the MIT wing of the Human Practices thrust has adopted this approach as well.

The critical limitations of such trained capacities and incapacities, however, are evident in their general outline, if not their specifics. First, the opportunity to develop an innovative and responsible post-genomic form will be blocked. Second, should things ramify into domains replete with dangers (known and unknown), there will be no internal mode of operation that would be capable of turning those dangers into risks that might be assessed in such a way as to manage them in an optimal manner. Third, given this situation, should security events take place, preparedness actions will come from governmental authorities whose power far and away trumps the minor sovereigns of the biosciences. Fourth, there is a distinct likelihood that a Monsanto-like event and reaction will take place, in which the framing of synthetic biology—first discursively and later regulatorily—will be given over to external critics.

The manifesto stage is relatively enjoyable. The trained capacity/incapacity of filling in the boxes in an already existing spreadsheet is painless, if time-consuming, and apparently some find it rewarding and are routinely rewarded for their diligence. It is true that facing up to the fact that manifestos are not research programs does slow things down, but it equally demonstrates seriousness, diligence, and discipline. *Observing and analyzing* the motion of ramifying pathways and forms in something approaching real time are conceptual challenges, with their own pleasures. Currently, however, how to address the challenge of inventing a form to collaborate in *inflecting* ramifications in something like real time, within a situation of indeterminate pathways and unequal power relations, is an unknown. Solving unknowns in such a venue requires thinking as well as the capacity to intervene. By the end of 2008, we have made real progress on the former and have been blocked on the latter. As of now the biosciences are moving into post-genomic territory accompanied by a set of human scientific correlative categories, practices, and (in)capacities from a prior genomic formation.

5 Lessons Learned 2009

From Discordancy to Indeterminacy

As of 2009, our experimental results indicated that in important, yet ultimately constructive ways, the initial design of the Human Practices thrust at SynBERC was proving to be extremely difficult to implement. These results were nevertheless ultimately constructive because experimentation is undertaken to discover something about reality, and what counts most in experimental terms is the contours of that reality, not one's hopes or desires. In that light, the results were neither disastrous nor inchoate: lessons can be drawn as to how the experiment unfolded and, given its initial objectives, went awry. Consequently, the purpose in pursuing this analytic work is to provide a reflective assessment that should make it possible to proceed with a rectified experimental practice. We pose the question to ourselves: What are the adjusted series of steps required in order to proceed toward a remediated compositional strategy that will better facilitate our experiments?

In this chapter we take up the task of identifying and arranging the elements that had been deployed in the first phase of the human practices experiment into two groups according to categories—*equipment and venue, contestation and secession*. The intent in adopting this approach is to facilitate a consideration of which of these elements might be decontextualized from the situation of indetermination or discordancy (or both) in which they had become embedded. To the degree that we can achieve this task analytically, the next step then would be to consider possible strategies of redesign of potential equipment through the recomposition of elements into modules. The task is remediative, transforming situational discordancy into discursive topics (i.e., a change of medium) so as to open the possibility of converting elements of our experiment into equipmental modules (i.e., to bring about an improvement).

Discordancy 1: Equipment and Venue

Although SynBERC was a new organization with ambitious goals oriented to building an innovative organizational form that would be suitable to scientific collaboration, we encountered an entrenched set of *equipment* and a desire to reinscribe an older *venue* within which much of that equipment had been originally designed to function. The source of both the equipment and the venue are easily traceable. The ease with which they were imported more or less whole cloth as well as the tenacity with which they were defended proved to be an unanticipated obstacle that we were not able to overcome.

Lesson: ELSI Is Dead; Long Live ELSI

There was a uniform adherence by each and every one of the participants in SynBERC to what we would characterize as either an ELSI model or an Asilomar model (or a combination of both). In either case, scientific and technical work is fundamentally cloistered, thereby guaranteeing a significant degree of autonomy for those engaged in that work. The price to be paid by the engineers and biologists for that autonomy appears to them to be minimal. Seen from the perspective of human practices, the price to be paid is extremely high. Currently, despite efforts by the NSF and other funders to insist on the inclusion of ethics as part of an overall research structure, little has been done by those in positions of power to guarantee movement in that direction.

- As discussed previously, the Asilomar model succeeded in framing all issues of security, participation, goals, and the like as safety issues. Elite bioscientists insisted that they were being responsible by instituting safety procedures. They acquiesced to temporary government advisory commissions. This set of relationships remains normative.
- The ELSI model introduced broad ethical topics into the cooperative framing but only on condition that the ethical, legal, and social topics were quarantined as "implications" and "consequences." Ethicists could establish bright lines delimiting zones of research and practices that should be considered illicit, but short of those bright lines, autonomy for the scientific practitioners was guaranteed. We found variants of this model in place.
- Both the Asilomar and the ELSI model structure interaction as cooperative. Both function so as to preclude collaboration.

• The Asilomar and ELSI models provide a rationale for a clear conscience for those adhering to them (bioscientists and human scientists
alike). They tell themselves that they are, after all, taking the necessary
steps to ensure safety; they are willingly complying with ethical limits.
And, as a surplus of morality and virtue, they encourage outreach and
a form of education known as vulgarization. The public needs to be as-
sured of safety, of ethical limits, and helped to understand, as best they
can given what they know and don't know, and instructed. And their
affect structure combined self-assurance and vehemence.

The stated agreement with the National Science Foundation was that we
would be supported in our efforts to invent a form of human practices that
integrated the valuable aspects of ELSI but moved beyond them. Otherwise
there was no reason to have an ethics component within the Center. We
found a uniform refusal to experiment or even entertain change. We concluded that old habits die hard. In and of itself, there is nothing surprising
about that. However, when old habits are combined with unequal power relations between those seeking to design a different venue and those seeking
to conserve an older one, then discordancy dominates. We failed to reverse
or even modulate this set of power relations. We did, however, receive support and protection from select members of the annual NSF site visit teams.
And outside observers and colleagues offered occasional sympathy.

Finally, both the Asilomar and the ELSI models were constructed with
and understand themselves predominantly within a *biopolitical* rationality.[1]
They take for granted that there is an object called "society." They take for
granted that there are probability series in place that can serve as the basis
for risk assessment. They take for granted that modern equipment will be
satisfactory for contemporary situations. They take for granted that ramifications can be reduced to consequences and implications. They take for
granted that there are available experts to perform these functions. We are
certain that each of these claims is contestable.

• The NSF constitution of its new centers required a commitment to
engage in active liaisons with industrial partners with the goal of fiscal self-sufficiency within a decade. This requirement was relatively
neglected during the first year of SynBERC's operations. During the
second year, it began to operate effectively and efficiently when a
director of Industry Liaison Relations was hired. His strengths, however, were also his weaknesses. He had worked in private industry for
the bulk of his career. While admitting that there were other valuable
things to undertake in scientific (and ethical) domains, his refrain

when presented with a paper or an argument was always "Will industry understand?"

- We proceeded from Rabinow's prior experience with certain perhaps atypical sectors of the biotech industry where there is a broader and more flexible understanding of goals and strategies than the purely instrumental.

- Over half of the NSF site review team for the second-year review worked in private industry. At a closing session where the team presented its initial SWOT (strengths, weaknesses, opportunities, and threats) bullet points, SynBERC administration underscored that one of the SynBERC test-beds—tumor-killing bacteria—did not meet standard industry criteria for specificity. One of us (Rabinow) intervened at one point to wonder out loud, "Given the hundreds of billions of dollars that the biotech and pharmaceutical industries had spent on research and development for anti-cancer therapies, and given the scanty progress in curing cancer that had been achieved, why should we not question existing industrial standards of specificity, research design, and the like?" His remarks were greeted with the habitual silence.

- Similar forces were at work during the experiment conducted during the summer and fall of 2007 to integrate a human practices practitioner into an iGEM team. Remarkably enough, although predictable, this type of collaboration had never been attempted before. The experiment turned out to be a rather painful event. The courageous Berkeley undergrad representing human practices was forced to address only instrumental considerations (intellectual property). The biopolitical and social consequences framing was brutally enforced to the extent that her slides were redone without her consent shortly before her presentation at iGEM. In a thoroughly unacceptable display of power, she was insulted in a preemptory manner in public by a PI who later apologized privately. Of course, given the circumstances, gender, and power relations, the rebuke should have been private and the apology public.

- The cost/benefit calculus put in place by ELSI can be formulated as follows: ELSI researchers, whether social scientists or ethicists, agreed to operate downstream and outside of the scientific research per se. The price to be paid for this positioning is that they remained in a cooperative rather than collaborative relation to the biosciences, and thereby could only contribute to their form at the level of after-the-fact regulations and the determination of "applications." The benefit of this positioning, however, was a relative rebalancing of power relations, on the model of earlier bioethical interventions into human practices research. This model establishes a governing body for vetting and thereby overseeing scientific research on human subjects. Science must justify its ethical standing by demonstrating that it is operating within prescribed moral boundaries, boundaries that are assumed not to derive from the field of scientific work, but are imposed from without. The ethical standing of science is guaranteed through the minimization of excess, not through reconstruction.

- The challenge is how to maintain the relative downstream balance of power between ethics and science characteristic of the ELSI model, while operating in an upstream or midstream collaborative mode. Perhaps these goals cannot be reconciled.

··

Diagnosis: Without a different set of power relations (enforced from the outside or from more enlightened leadership on the inside), the current SynBERC venue is an inappropriate one to carry out the experiments in human practices that we propose. The hope is that by adding other sites outside of SynBERC, we might be able to continue working in a modified and reduced capacity in the original venue.
··

Lesson: Start-up Principles—Founders Are Not Managers

It is a well-known maxim of start-up enterprises, drilled home by venture capitalists, that founders should not be able to serve as managers (or CEOs). The reasons that founders are good at founding enterprises is that they have big ideas, large visions, strong and well-protected egos, have received support and praise for long periods of time, and have often, but not always, accomplished impressive, if highly specific, scientific or technological breakthroughs. Founders tend to confuse their personal qualities and skills as well as the support and praise they have received throughout their careers with the qualities and skills required to transform their insights into research programs organized around standards of accountable progress toward instrumental goals. The very skills that launch a project may well be precisely those skills that can stand in the way of bringing it to fruition given the quotidian drudgery, incessant attention to the accuracy of microdetail, recurrent demand for compromise in goals, and the necessity of operating in an environment of sustained competitiveness. Such an environment encourages a degree of scrutiny to which founders are frequently unaccustomed.

As we have described, the pathways from manifestos to research programs can be more difficult to traverse than anticipated, especially by relatively inexperienced if acclaimed practitioners. In our experience with SynBERC (and the larger community of synthetic biologists and engineers), although traversing that threshold can prove difficult for particular individuals, it has not posed any especially major obstacle for the larger community of more and less advanced researchers who are not themselves committed to the details of the original manifesto. The community of synthetic biologists currently includes large numbers of undergraduates (iGEM), graduate students, postdocs, assistant professors, and so on, up the professional ladder. What one team does not achieve another may see as an opportunity for advancement and increased visibility in this competitive environment. As is typical in science, we have observed the pleasure and claims to higher status occasioned in saying that another team's view of, say, parts or devices is overrated or impractical, and then demonstrating how it could be done better. Better can mean less discursively neat but more appropriate to a specific experimental system.

One of the central characteristics of the way financial support has been apportioned in SynBERC is its egalitarianism. In the first few years, each principal investigator received a fixed sum according to a budget formula. One of the main unintended ramifications of this formula was that no lab (with the exception of the Human Practices labs) received enough money from SynBERC to support their activities in full or even in a major way. Hence the commitment of the PIs to the Center was based in somewhat equal parts on goodwill as well as their investment in the growth and success of the field and brand—based in part, that is, on a calculation of the benefits (scientific, technological, logistic, affective, symbolic, etc.) that can be derived from an often minimal cooperation. The flip side of this arrangement was that none of the labs involved in SynBERC really depend on it (again with the exception of the Human Practices labs). When these structural conditions of low finance/low commitment were combined with Keasling's hands-off management style and temperament, as well as the unequal power relations in regard to Human Practices and the apparent willingness to reinforce those unequal relations both tacitly and actively—one can only conclude that major organizational changes are not on the agenda.

- Founders who had come to prominence by framing all aspects of the project resisted (and at times attempted to sabotage) the efforts of the Berkeley Human Practices Lab to provide a framing that was different than their own both in content and in its scientific legitimacy.
- The basic tactic deployed widely within SynBERC was to delegitimate Berkeley Human Practices and to proclaim that it produced observations and insights that were not comprehensible. Blackberry in hand, and airline connections foremost in mind, such critics neglected to add

that they did not consider it worth the investment of time or thought to learn new terms and analytic approaches. When pedagogy or collaboration requires change, the patterned response has been to assert dominance.

• The defense of their (unreflected-upon) biopolitical framing of social consequences, benefits to the population, normalization, and accelerated circulation served to reinforce the technocratic and charismatic formulas.

• Although Human Practices materials and publications have been available to anyone who wanted to have them, no substantive responses, including from MIT Human Practices, were offered.

Diagnosis: A venue in which founders retain management power is unlikely to progress toward organizational innovation, agility, and accountability. Without the functional equivalent of the venture capital overseers vigorously cajoling all involved to produce results or to adhere to a common set of audit standards, as would be the case in an industrial setting, little is likely to change.

Discordancy 2: Contestation and Secession

Faced with the obstacles of the entrenched (unrecognized and cathected) *equipment* and *venue*, the realization began to solidify that we were incapable of overcoming trained incapacities, or even making them topics of serious discussion. Human practices, whatever else we would like it to be, must provide conceptual analysis of emergent problems, so as to specify and reflect on their scientific and ethical significance. Just as biologists are trained to be alert to what is significant in bioscientific results, a task of human practices, as we are trying to design it, involves techniques of discernment and analysis that facilitate collaboration by specifying and characterizing new problems. This is precisely where Mode 3 Human Practices departs from Mode 1. And this is precisely where our work was most assertively blocked.

We have been unable to make any serious progress in working with the SynBERC engineers and biologists to define what counts as significant Human Practices problems. Our multiple attempts to formulate and reformulate a core set of problems were almost invariably met with the demand that Human Practices concentrate on issues that biologists and engineers bring to the table, typically matters of intellectual property or public relations. Even in ELSI, whatever its other limitations, ethicists and human scientists were not put in the position of being consultants whose job is simply to provide justifications for the biologists and engineers.

As we experienced our efforts being blocked, it became clear that ethically and scientifically we were going to have to attempt something different. Active attempts to explain our approach inevitably devolved into demands that we justify the worth of our efforts. A different form of contestation was called for. Given an available repertoire of concepts addressing the subject position of the anthropological observer, as well as those addressing power relations and networks, the time had come to consider alternate paths of action.

The development economist Albert Hirschman published an insightful book in 1970 entitled *Exit, Voice, and Loyalty: Responses to Decline in Firms, Organizations, and States.*[2] His characterization of strategies available to deal with "decline" or "blocked" situations is helpful in identifying the choices and options we had been considering as a pathway out of this discordant and unproductive situation. We added a fourth alternative to Hirschman's list.

Lesson: Adjacent Means Proximate, Not Internal

Given the aforementioned conditions of habits, structures, rationalities, and power relations, we had to begin developing a diagnosis of what was going on, and then design and follow with a response that proceeded from that diagnosis. That response had to balance affective, tactical, and strategic components. When it was clear that cooperation with the MIT Human Practices team was not going to be fruitful or even possible, we proposed a solution where we would have separated work into two distinct and basically autonomous approaches: Berkeley would be responsible for one set of issues (ethics, institutional innovation, and ontology) and MIT, another (intellectual property, policy assessment, and risk analysis). There would be a third cluster of issues around biosafety, biosecurity, and bio-preparedness that we would share. This division of labor was filibustered by Oye. And although it was eventually accepted by Keasling and made official, it was not given the kind of support that would have made it effective.

These responses were especially frustrating as there was by the fall of 2007 an incessant demand to produce materials for the report to the NSF that would structure their second-year site visit. Oye wrote a large number of e-mails, left many phone messages, but produced little or no prose that fit the requirements. It eventually fell to one of us (Bennett) to draft most of the material, a task that took close to a month. Oye objected again to a draft version that had met with approval in Berkeley, but he provided no specific alternatives. Finally, Keasling accepted our version verbatim but said that it would be best to tell Oye that he, Keasling, had written it. The Orwellian injustice of the whole situation was brought home to us when, in one of his whirlwind visits to Berkeley, one of the SynBERC PIs said that he had written his section of the report the night before, "almost an all-nighter."

There were vague suggestions floated during this period of time that Rabinow should be made head of Thrust 4; suggestions that were not implemented for another year. And there was repeated agreement that Oye was not producing his contractual first-order deliverables. Whether the failure to resolve the situation in a productive and satisfactory manner was a tactic to keep us involved and working or not, we do not know. What we do know is that we were faced with an unproductive and deadlocked situation—a situation in which we did not have the possibility of putting into practice the kind of equipment we had designed. In sum, we were faced with a situation in which we saw no conditions that would lead to our own flourishing, not to mention our by now grandiose-sounding designs for a broader flourishing.

···

Diagnosis: We are experimenting with an *adjacent* position. *Adjacent* is defined as "situated near or close to something or each other, especially without touching." This positionality has been conceptualized as pertinent and salient for the anthropology of the contemporary.* There is no question that if the goal was to write an anthropological account of SynBERC, this positionality would be ideal. However, the goal has always been something different than that: the design and implementation of collaborative equipment that could contribute to flourishing. Whether this stance of adjacency to SynBERC and in a proximate fashion to other centers will prove feasible remains to be tested.

···

*Paul Rabinow, *Marking Time: On the Anthropology of the Contemporary* (Princeton, NJ: Princeton University Press, 2007).

Lesson: Bite the Hand That Feeds You

We had attempted various forms of intervention, and all were met with the same response: Do whatever you want to do as long as it does not cause trouble in public and as long as it does not oblige us to spend significant time or effort in accommodating it —"That is what all the other PIs do, why don't you do it?" One might say that here "loyalty" was proposed, and the price to be paid was a certain bad faith—but a bad faith that would allow us to continue the broad outlines of our work, albeit not within SynBERC directly.

We were faced with a variant of the dilemma Hirschman describes of how to respond to a blocked situation. In Deweyan terms, such a response required more thought, by which he meant a reflective practice that intervenes in an indeterminate or discordant situation. But Dewey had little or no place in his tool kit for accounting for unequal power relations and ethical unconcern. Consequently, we considered resigning. Aside from the pride (in Aristotle's sense of a free citizen not voluntarily acquiescing to insults) involved in not giving in to an unjust solution, we were not sure what such a gesture would accomplish. It took months of hesitation and irritation before we agreed that we could not continue to be stymied and compromised in our scientific life and ethical commitments, regardless of the fiscal price to be paid.

Put more positively, the contract proposed appeared to put us in a rather traditional anthropological position of observer. We could remain as participant-observers as long as our participation was strictly limited and followed a rationality dictated by others and conforming to a mode we were trying to redesign.

- This condition of compromise and stasis reigned for the first two years. Increasingly as time wore on, the requests to provide materials that would be useful to "the scientists and to industry" became more pronounced and insistent. We produced a diagnostic overview of synthetic biology (www.bios-technika.net) that differed significantly from the manifestos that had accompanied the birth of the brand. It was received positively as a useful contribution and was used by SynBERC in presentations to industry. It was deemed to be comprehensible to outside audiences and that the shift away from the original manifesto language was getting much better results. But this contribution on our part seems to have been soon forgotten, and new demands of the same type arose.
- As far as we know, none of these demands to be useful to industry were made on Oye. His work on intellectual property and the idea that there would be an open-source patenting policy was held to be pertinent to some and irrelevant to others. The industrial liaison officer insisted that no one in industry would entertain such a policy.

Having considered the pros and cons of "exit" and "loyalty," the remaining response was "voice." We had voiced our concerns and our insights internally. During 2007 we devoted some effort to establishing external relations through which we could give voice to our understanding of how synthetic biology in general and SynBERC in particular were ramifying.

- During the latter part of 2007 and into 2008, one of us (Rabinow) was contacted by a growing list of media (*Washington Post, San Jose Mercury News, Baltimore Sun*, Canadian Broadcasting Corporation, National Public Radio's *Science Friday*, etc.). We were still operating with a "loyalty" mode and attempted to be careful to find a manner of describing projects and ramifications that would support the broad goals of the enterprise without contributing to a mood of fear about the ramifications or consequences of cutting-edge science and technology. Hence we voiced no overt denunciatory criticism in public. However, even this mode produced a negative reaction.
- In addition to regular interactions with the media, the Berkeley Human Practices Lab developed relationships with parallel organizational efforts under way in other NSF-funded centers, specifically nanotechnology. The NSF had funded thirty such centers with a federal mandate to cultivate the field, to encourage strong relations with industry, and to include a robust social consequences or ELSI component. The Center for Nanotechnology in Society at Arizona State University had the largest and most thought through Science and Society component. We made contact with them and were well received.

The Year Two and Three NSF Site Team Reports, while not proposing any organizational or venue change for SynBERC as a whole or the Human Practices thrust as a part, were very supportive of the Berkeley Human Practices Lab's efforts to "redefine the task and deliverables of ethics research."[3] As we were quite vocal with our overall criticisms during the site visits, perhaps growling at those who feed you can be productive in focusing their attention.

....

Diagnosis: At this juncture, we are attempting to juggle "voice," "loyalty," and "adjacency." We are experimenting with broadening the framework within which issues of flourishing can be taken up. We are still entertaining the possibility that this adjacent position can provide a pathway in and out of SynBERC. In sum, we are attempting to "secede," that is to say, to avoid direct confrontation or simple exit by realigning a number of other sites into what might become a venue that could facilitate the experimentation with distributing aspects of our work in different sites so as to produce a form that might accommodate the significant categories and elements of our overall program.

....

Significantly after the NSF site visit for year three, we concluded that we had moved from a situation of indeterminacy to a situation of determinacy, but that those determinations were, given our objectives, discordant ones.

Again following Dewey, it is worth remembering that a discordant situation is one ripe for inquiry and thinking. In that light, we can now identify short-comings in skill on our part as well as both passive and active resistance to our initial efforts on the part of others. In the best case, most skillfully handled, the lessons learned can function as topics opening up pathways that might prove fruitful for the design and composition of different and better forms of experimentation.

An unsuccessful experiment is not a failed experiment. An unsuccess-ful experiment can be considered worthwhile for several reasons. It might demonstrate that the whole experiment was misconceived; there was no real *problem* to which the experiment could provide answers or clarifica-tions. It might demonstrate that the experiment was crude in its initial formulation and needed to be refined so as to become fruitful. It might demonstrate that the nature of the *objects* (and their component *elements*) under consideration were incommensurate with the larger goals of under-standing that the experiment was designed to explore in the first place. It might demonstrate that aspects of the experiment needed to be modified so as to *perform* it better. In the latter case, if one can identify the elements that functioned as expected and those that did not, then one would have made scientific progress. And one would then be in a position to recast the experiment in more precise *terms*.

More specifically, undertaking an analysis can indicate mistakes in the handling of materials, reagents, and technical procedures. It points to the degree of *skill* or *incapacities* of the researchers, thereby opening up path-ways for renewed *training*, *exercise*, and *remediation* of the *practices* involved in the experiment. Finally, it might indicate that regardless of how the given elements, objects, procedures, and skills were handled, the *venue* was in-appropriate and, regardless of the degree of competence involved in carry-ing out the experiment, the results produced would be by definition incon-clusive as to the status of the problem. In that case, one would need to focus on finding or inventing a different form and venue for the experiment.

PART II

Human Practices: Inquiry

6 Recapitulation and Reorientation 2009

The First Wave of Human Practices

> It is not the "actual" interconnections of "things" but the conceptual interconnections of problems which define the scope of the various sciences. A new "science" emerges where new problems are pursued by new methods and truths are thereby discovered which open up significant new points of view.
>
> MAX WEBER[1]

By three years into our initiative, it was apparent that SynBERC PIs had contributed to establishing synthetic biology as a leading post-genomic brand and stabilizing synthetic biology as a (heterogeneous) research community. It was also clear that human practices, broadly understood, had become a feature of both that brand and to a lesser degree of those communities. A primary factor for this inclusion has been the availability of funding, institutional support, and professional recognition for human and social scientists to work on social consequences, public opinion, and ethical implications. Even with support and recognition, however, cooperation has remained slack.

As of 2010, major human and social sciences organizations and players in Europe and the United States had initiated multiple projects. In the United States, for example, the NSF was supporting the Human Practices component of SynBERC and the Alfred P. Sloan Foundation was supporting coordinated projects on security at the J. Craig Venter Institute,[2] bioethics at the Hastings Center,[3] and public and policy discourse at the Woodrow Wilson International Center for Scholars.[4] The European Council has funded multiple projects, including the Synbiosafe project coordinated by the Organisation for International Dialogue and Conflict Management and involving organizations in Austria, Switzerland, and France.[5] The UK's Biotechnology and Biological Sciences Research Council (BBSRC), funds the Networks in Synthetic Biology project, which includes science studies researchers at Oxford, Nottingham, Cambridge, and Edinburgh, among others.[6] And Euro-SYNBIO is entering a first round of funding for "societal" research, having received proposals for coordinated research among scholars in Switzerland, Belgium, Germany, Austria, the UK, and the United States (a consortium of which the Human Practices thrust at SynBERC is a proposed partner).[7]

In 2009 an article by two bioengineers from a leading lab, Priscilla Purnick and Ron Weiss, came to our attention. Entitled "The Second Wave of

Synthetic Biology: From Modules to Systems," the article provides an overview of the bioengineering state of play in synthetic biology (we will turn to
the details of this article in the next chapter).[8] Purnick and Weiss's analysis
of the limitations of current practices and of the need for more effective design strategies, as well as their term "second wave," is astute and judicious.

The article triggered a question for us: What would qualify as a second
wave of human practices? In order to respond to this question, we decided
that we first needed to clarify whether there had been a broadly coherent
first wave of human practices for synthetic biology. If so, what were its defining traits, and what have been some of the key lessons learned during
this first period? We should emphasize, in contrast to Purnick and Weiss's
account of the second wave of bioengineering, we do not think a second
wave of human practices yet exists in a settled form for synthetic biology,
although we are convinced that most of the main actors involved have indicated on their own the need for something like it. Our goal in this chapter is
to specify questions, challenges, and limits that have characterized the first
wave of human practices. In this light, we will propose appropriate initial
parameters and modes of engagement for human practices' second wave.

Broadly speaking, first-wave work has focused on *diagnosis*. Common
questions have been: Does synthetic biology pose new problems? Are current disciplinary forms and infrastructures sufficient? Claims that "social
and ethical concerns" need to be posed are consented to by almost everyone
involved, and most social scientists offer arguments justifying some mode
of "upstream" engagement. What each of these subject positions takes to
be the principal object of concern, what kind of knowledge they offer, what
form of intervention they recommend, and which venues they operate
within, however, vary considerably. Working schematically and heuristi-
cally, we have produced a diagram that distinguishes four first-wave subject
positions: (1) policy, (2) bioethics, (3) science and technology studies, and
(4) human practices (see table 6.1).[9]

First-Wave Diagnosis: Does Synthetic Biology Pose New Problems?

Subject Position	Security Policy	Bioethics	STS	Human Practices
Object	Security/Expert Opinion	Social Issues	Actors/Networks	Figures/Ontology
Mode of Veridiction	Survey	Alignment	Mapping	Warranted Assertion
Mode of Jurisdiction	R&D	Regulation	Critique	Remediation
Venue	Governmental Initiative	Ethics Center	Network	Multidisciplinary Center

As our goal here is to open a discussion of second-wave practices and problems, we do not provide a detailed discussion of first-wave projects; furthermore, much of the first-wave work itself takes as its task presenting a comprehensive overview of the sectors with which it is primarily concerned. Rather, we adopt a second-order position in regard to these first-wave proposals (including our own). We seek simply to identify what we take to be their strengths and limitations.

First-Wave Security Policy

A consistent expectation from the funders of synthetic biology is that the principal focus of human practices research should be on matters of concern grouped loosely and somewhat confusedly under the term *security*. Government officials and advisory commissions and committees stressed that the hoped-for outcome of such a focus would consist of policy guidelines for the regulation of synthetic biology in such a way as to minimize adverse security events.[10]

To date, work on security policy has been framed in terms of the challenge of risk assessment. What kinds and levels of risks does synthetic biology present, and what kinds of regulations can be put in place in view of those (postulated) risks? A central—though insufficiently examined—supposition of policy experts and other concerned actors is that they are capable of conceptualizing dangers, both known and unknown, as risks that can subsequently be assessed with appropriate rigor and plausibility. The existence of such a capability is very much in question.[11]

Two of the first-wave projects exemplify efforts to meet this challenge. The first was directed by Stephen Maurer, a lawyer with strong interests in economics and biodefense. Maurer proposed, and argued forcefully for, two things: a mechanism to monitor "experiments of concern" and a procedure whereby the "community" of synthetic biologists would vote on a set of regulatory controls that would govern the relations of the nascent DNA synthesis industry and the "community" of synthetic biologists. The substance of Maurer's proposals was eventually developed in a report funded by the Sloan Foundation.[12]

Maurer's initial work was circulated in a report published by the UC Berkeley Goldman School of Public Policy.[13] The report drew attention to the need to monitor the solicitation of potentially pathogenic DNA sequences from commercial firms. It was proposed that such sequences could be identified by as yet to be developed database of potentially dangerous sequences, procedures for screening, and software to facilitate processing. The report is proceduralist and advisory in tone and emphasis. It draws on the 1975

Asilomar convention as a normative model for synthetic biology, pending appropriate updates. The report argues that community governance, rather than external regulation, is likely to be most advantageous for scientific, technical, and security development.

Another of the first-wave projects on security policy is that directed by Professor Alexander Kelle in a working paper written for the European Union's Synbiosafe research initiative.[14] One key aim of the Synbiosafe initiative is to determine the relative conscientiousness (or awareness) of biologists relative to questions of security and synthetic biology. Based on responses from 1,600 life scientists during sixty seminars in eight countries, Kelle concludes that life scientists "do not share the threat perception widespread among biosecurity experts concerning bioterrorism or biological warfare. They do not think that their own work might contribute to the threat. Life scientists have practically no knowledge of legally binding international regulatory instruments, such as the Biological Weapons Convention."[15] Presumably, increasing the number of biologists who do take these issues seriously will decrease the likelihood of a dangerous event occurring, although such likelihood—to make the point again—still could not be calculated under traditional models of risk assessment.

The strength of such initial policy efforts is that they have provided an orientation to key challenges and have taken preliminary steps toward providing diagnostic orientations to the elaboration of policy. Thus, Maurer et al.'s advance is that they identified the need to provide an initial plan for screening protocols and pressed (albeit as yet unsuccessful) efforts to formalize a community of self-regulation. Kelle et al.'s advance is that they provided empirical confirmation of the resistance or simple disinterest of the overwhelming majority of biologists and engineers to engage questions of security.

Both projects take a *communicative rationality* approach to the challenge of formulating policy guidelines.[16] In such an approach, what counts as a fact and a truth claim are opinions, whether expert or popular. Thus, having interviewed scientists, both reports conclude that synthetic biology raises no fundamentally new security challenges, while recognizing that the goal of making biology easier to engineer certainly can be seen to amplify existing challenges. Thus, in another study, the Synbiosafe group cites interviewees who conclude that the only significant new concerns raised by synthetic biology "would be those that arise from public fears, or the 'Frankenstein Factor.'" They do not, however, foreclose the emergence of future developments, and propose hosting a series of conferences and forums as a "favorable precondition for a constructive exchange of ideas and the ability to achieve rational, multi-stakeholder governance of the discipline."[17]

Second-Order Observations

First-wave policy approaches combine two familiar modes of social science engagement with the life sciences, exemplifying the strengths and limitations of each. In surveying the synthetic biologists, they take an approach that we have called "Mode 1: Representing Modern Experts." Their strategy is to counter claims that synthetic biology introduces new security issues by appealing to "the views of the scientific community itself."[18] Aside from the vested interests of those being interviewed, and the question of whether or not expertise in biology and engineering qualifies scientists to speak authoritatively on matters of security or international politics, the major limitation of this opinion survey approach is that in an emerging domain such as synthetic biology, adequate expertise—for problems not yet formed—does not exist.

In complement and supplement to the limitations of Mode 1, Synbiosafe researchers hosted an e-conference and have proposed additional multi-stakeholder conferences and forums, perhaps another Asilomar. This supplement exemplifies what we call, following Nowotny et al., "Mode 2: Facilitating Relations between Science and Society."[19] Mode 2 has become the norm for policy in Europe and is beginning to be institutionalized in the United States. The strength of Mode 2 is that it recognizes the limitations of existing expertise and addresses the inequality in power between the biosciences and other disciplines. It does this by engaging a broader set of stakeholders in research and development. Its limitation is that strategies for identifying society's representatives are frequently contestable by those who consider themselves to be excluded or have never heard about the possibility of being represented. Moreover, as our work and the work of others have shown, opinion polling, proceduralism, consensus building, and the multiplication of representatives' expression simply cannot adequately address the ethical and political question of whether or not a given course of action is good or bad, right or wrong, just or unjust, regardless of how many people favor or oppose it in a survey or forum.

Consequently, we can (partially) concur with the Synbiosafe scholars' assertion that "whereas it may well take some time to fully grasp the philosophical challenges that synthetic biology presents, it will be important to create the necessary space for an informed, participatory discourse to accompany the development of this discipline."[20] We would add, however, that such a participatory discourse, in and of itself, is unlikely to answer the question of the distinctiveness of the current conditions within which synthetic biology is ramifying and to which it is contributing. Although Maurer and Kelle succeed at representing experts and facilitating dialogue, we underline that a further challenge not addressable by these methods consists in characterizing and assessing the current empirical conditions within which research and development is being elaborated as a means of specifying and evaluating relevant problems.

First-Wave Bioethics

A number of prominent bioethicists have asked whether or not, or to what extent, technical and organizational innovations in synthetic biology call for parallel innovations in bioethics. The answer has been taken to turn on whether or not synthetic biology raises any new *ethical issues*.[21] These explorations have been constructive insofar as they set the stage for conducting inquiry into an important problem: Is American bioethics, in its current forms, adequate to developments in post-genomic biology generally and synthetic biology in particular as well as in the complex risk-filled twenty-first-century world?

Professor Laurie Zoloth of the Center for Bioethics, Science and Society at Northwestern University was the first bioethicist to give sustained attention to the question of synthetic biology and bioethics. She oriented her work through a careful parsing of old and new issues. She has argued that if synthetic biology is taken to be only a variant on genetic engineering, then, unsurprisingly, many of the issues addressed by bioethics in other contexts remain pertinent. Relevant difficulties, in such a case, will basically concern regulation: "Who should regulate these sorts of trials, which fall outside traditional categories of genetic manipulation, human or animal subjects, or dangerous materials?"[22]

Zoloth, however, accepts the argument made by some biologists and engineers that synthetic biology is not simply a next generation of genetic engineering. The primary objects of interest, after all, are not the molecule or the gene but the "part," "device," or "chassis," to use just one prominent framing. Synthetic biology engages long-standing concerns, to be sure. Among these Zoloth specially notes "questions about 'naturalness,' scientific hubris, error, and safety." But, as Zoloth rightly argues, these familiar concerns are not decisive for determining synthetic biology's ethical distinctiveness. Crucial is the fact that synthetic biology "came of age at a particular time, after the rise of global terrorism and the emergence of new diseases with lethal epidemic potential."[23] Moreover, unlike "traditional" genetic engineering, many synthetic biologists are attempting to design and construct *de novo* living organisms. These efforts elicit "special issues." Zoloth asks whether attempts to incorporate evolution's design principles and processes (e.g., mutation, random selection, and transformation, etc.)

into an engineering tool kit could be regulated at all.[24] With the possible exception of questions of safety, which themselves are not at all obvious, on what grounds could the effort to bring new things into the world be regulated? In light of this difficulty, Zoloth proposes that as synthetic biology matures, bioethical guidelines be developed less with an eye toward familiar regulatory techniques and more with "deeply informed" attention given to the new things that synthetic biology is bringing into the world. Said another way, Zoloth highlights the significance of future ontological questions.

Erik Parens et al. of the Hastings Center, one of the long-established centers for work on bioethics, take up the question of whether or not synthetic biology raises any new ethical issues by posing a related, but different question: Is it a good thing to proliferate subfields and subspecialists in the profession of ethics in a fashion parallel to the emergence of each new scientific subdiscipline?[25] Framed this way, we concur that the answer is almost certainly no. Following a precedent that has been characteristic of American bioethical responses to developments in genetic engineering since 1979,[26] Parens et al. proceed by distinguishing questions of safety and security from questions of meaning and social well-being. Among other things, this distinction works to demarcate a space of expertise and competence for professional bioethicists alongside other subject positions, such as law, risk assessment, and policy.

Having marked out this space of appropriate competence, they can then take as their primary task the work of aligning and distinguishing old and new ethical issues. Framed this way, a central task is to discern the extent to which synthetic biology raises new issues or simply reactivates, albeit in new contexts and with new capacities, issues that bioethicists have already sorted through. Parens et al. conclude that synthetic biology does not raise new issues, though it may reactivate and catalyze variants of established concerns. They argue therefore that it follows that new ethical subdisciplines are not called for currently.[27] Rather, their proposal is to adjust existing bioethical apparatus such that "familiar questions" can be taken up in relation to "new scientific contexts."[28]

Second-Order Observation

The advance offered by first-wave bioethics is that it has aligned existing ethical issues with developments in synthetic biology so as to parse old and new questions. In our view, however, posing the question of ethics and synthetic biology as a matter of old and new issues deemphasizes or obscures other significant considerations. Significantly, this formulation covers over the fact that American bioethics emerged in a specific historical and national context; it was not the discovery of timeless universals but a particular response to a distinct set of problems.[29] Parens et al. would no doubt accept this historical fact, and Zoloth, for her part, makes a point of underscoring it. To date, however, bioethicists engaged with synthetic biology have not pressed what seems to us to be a crucial line of inquiry: Is there something distinctive about contextual changes in the life sciences today, the venues in which post-genomic biology is being invented, and the modes in which it is ramifying into other domains (security, global networks, environmental problems, etc.) that warrants a reexamination not only of these specific topics but more generally of reigning bioethical orientations and practices?

Historically, the signal achievement of American bioethics was its development of practices, procedures, and principles calibrated to specific problems (protection of human subjects in research, issues of justice, the need for bureaucratic norms for health care, etc.). The founders of American bioethics were keenly aware that this calibration of a mode of ethics and problems, in turn, entailed the construction of specific new venues (e.g., IRBs), distinct modes of collaboration (e.g., advisory government commissions), and particular types of inquiry (e.g., the rise of bioethics as a discipline). Today it seems not only appropriate—but scientifically and ethically mandatory—to consider whether these bioethical practices, procedures, and principles remain adequate to current conditions, and if not, how they might be remediated?

Given our experiences at SynBERC, we can partially concur with Parens et al.'s assertion that "when bioethicists think they have found a new set of ethical questions, they are prone to think they can provide a new set of answers."[30] We, however, would ramify this assertion in a different direction. Where Parens et al. see a primary task as consisting in parsing old and new questions, we conclude that the first challenge consists in characterizing and assessing the current empirical conditions within which research and development are being elaborated. At issue is not only a change of discourse but pragmatic attention to venues and practices: How things come into existence—are named, sustained, distributed, and modified—is an issue of primary importance for addressing ontological, ethical, and regulatory questions.

We concur with Zoloth that the question of old and new *ethical* issues cannot be answered without posing *ontological* questions. Whatever else it does, synthetic biology makes new things. What new types of regulation would be practical and effective against possible new biological entities and capacities is a widely shared concern. What is missing from such questioning, in our view, is sufficient evidence from empirically based human sciences inquiry into such questions. How things come into existence—are named, sustained, distributed, and modified—is an issue of primary importance.

First-Wave Science and Technology Studies

We have observed that in the first wave of human practices, policy work has surveyed key challenges and taken initial steps toward providing important diagnostic openings. We have also observed that these preliminary steps nonetheless neither provide a sufficient security barrier for nefarious or unexpected uses of synthetic biology nor do they provide grounds beyond safety considerations for normative guidelines. Equally, we have observed that the first wave of work in bioethics has aligned existing ethical issues with developments in synthetic biology so as to parse old and new questions. We have also observed that these initial steps nonetheless have not yet taken sufficient account of the broader contemporary context in such a way as to press the question of the sufficiency of previous bioethical apparatuses.

We now turn to the third subject position operating in the first wave of human practices. The strength of the initial STS efforts concerning synthetic biology is their empirical emphasis on mapping emergent actors and networks. Andrew Balmer and Paul Martin of the Institute for Science and Society at the University of Nottingham were the first to provide an initial map of topics, problem, areas, and key actors. Their advance is that they demonstrated the significance of the interfaces among these diverse elements.

Maureen O'Malley, Alexander Powell, Jonathan F. Davies, and Jane Calvert provide initial mapping of strategic epistemological distinctions across synthetic biology research programs, albeit distinctions not uniformly accepted by the biologists and engineers. Their contribution is to advance hypotheses about possible significant epistemological fault lines between and among competing groups within and beyond synthetic biology. O'Malley et al.'s major contribution after their epistemological mapping is a critical observation. They argue that synthetic biology needs to attend carefully to the underlying biological assumptions that inform its research strategies.

Second-Order Observation

We concur that the STS contributions, especially those of O'Malley et al., frame synthetic biology in a critical manner that highlights the limits of current research strategies concentrating attention on the pragmatic significance of epistemologies. Although this mapping is helpful, verifying its claims will depend on close observation of specific labs and research programs as they unfold. We fully agree that engineering analogies can be misleading and that careful attention needs to be paid to the character of biological systems. Analogies, however, can also be misleading if they presume to prescribe unavoidable limitations in advance and a priori.

Having spent several years embedded in a synthetic biology research center, we have found that overly strong claims as to the limitations of engineering analogies should be tempered through close observation of design strategies and dialogue with researchers in the lab. Our anthropological embeddedness has made us aware that many of these critical issues are in fact being confronted in an experimental, if frequently ad hoc, manner by leading researchers, albeit in a less social scientific manner. At issue, once again, is not only attention to framing discourse but rather pragmatic engagement with venues and practices: How should we rethink and rework reigning habits, dispositions, and power relations among and between the life sciences and the human sciences?

First-Wave Human Practices

In a similar manner to the other first-wave subject positions, our initial work in human practices has proceeded in a diagnostic mode: specifying and evaluating relevant problems and proposing appropriate concepts. In contrast to other first-wave subject positions, however, we did not enter into this domain as part of an existing disciplinary apparatus whose expertise was authorized and settled. Rather, from the outset our orientation was pragmatic and experimental.

During the first wave of our work, we framed the challenge as follows: To what extent might an adjusted mode of anthropological inquiry facilitate the challenge of rethinking and eventually putting into practice some form of a "post-ELSI" program for synthetic biology? We began this work intending to produce a diagnosis of a new "problematization" taking shape in the world. Although the contours of what seemed to be emerging were vague, we had a strong sense arising from a great deal of discussion, analysis, seminar work, and reading, that whatever was taking shape could not be sufficiently characterized by reigning analytic doxa.[31] Our commitment and orienting challenge was to develop strategies for research, collaboration, and dissemination, which bring together biological, anthropological, and ethical understandings of synthetic biology. We argued that conceptual and design strategies needed to be "upstream" and/or "midstream" in their execution, and that work needed to be done producing deliverables whose form is consistent with the dynamic and evolving character of the biological research proposals.[32]

Instead of focusing on technical innovations, the parsing of old and new issues, or the mapping of existing actors and networks as a means of determining relevant problems, our strategy was to proceed by posing the question: How is it that one does or does not flourish as a researcher, as a citizen, and as a human being? Flourishing here involves more than success in achieving projects; it extends to the kind of human being one is personally, vocationally, and communally. Contemporary scientists, we repeat, whether their initial dispositions incline them in this direction or not, actually have no other option but to be engaged with multiple other practitioners from disciplines and life-worlds quite different from their own. The only question is this: How best to engage, not whether one will engage.

The first wave of our experiment at SynBERC has confirmed that a primary challenge of constituting an appropriate relationship between and among human sciences and post-genomics consists first in analyzing current conditions through empirical attention to existing and emerging *ontology* and *figures*.[33] We were convinced that a careful diagnosis of ontology (i.e., what things exist and how they are made into objects) and figures (i.e., how things and objects in the world are framed and connected) would orient us to significant new problems and scientific approaches. Said another way, following Max Weber, we shifted our attention from the attempt to characterize the "actual interconnections of things" to an attempt to distinguish "the conceptual interconnections of problems," with the hope that we would be "opening up significant new points of view." Such points of view, we came to think, would be significant to the degree that we could transform these perspectives into actual practices.

It follows that, in the light of the results of future inquiry, it should become possible to specify the extent to which synthetic biology gives rise to distinctive and significant problems. Our first-wave advance was to provide a set of conceptual tools adequate for an analysis of the problem-space within which synthetic biology is emerging so as to reflect in a rigorous fashion on its ethical significance and ontological status. We proceeded by posing the question of the distinctiveness of the current conditions within which synthetic biology is ramifying and to which it is contributing. We proposed to diagnosis in a systematic fashion existing modes of engagement between the life sciences and human sciences (i.e., Mode 1 and Mode 2) as the initial phase of our work. The defining goal was to identify design parameters for what we termed Mode 3. Mode 3 shares with Mode 1 and Mode 2 a sustained concern for conceptual and practical work on the question of how to bring biosciences and the human sciences into a mutually collaborative and enriching relationship. Furthermore, Mode 3 takes head-on the fact that increases in efficiency or maximalization, or opinion polling, while often valuable, can never themselves provide a sufficient ethical metric. We adopted the term *remediation* as the challenge of both improvement and change of media. It followed that the challenge and task was to conceive of and design forms within which such integration and collaboration could prove possible.

Second-Order Observation

We have concluded that in important, yet ultimately constructive ways, the results of our first wave of experiments in human practices at SynBERC proved to be limited, though productive and revealing. The results were ultimately constructive, however, because experimentation is undertaken to discover something about reality, and what counts most in experimental terms is the contours of that reality, not one's hopes or desires. In that light, we emphasize again that the task is to draw lessons as to how the experiment unfolded and, given its initial objectives, was insufficient. Consequently, the purpose in pursuing this analytic work is to provide a reflective assessment that should make it possible to proceed with a rectified experimental practice.

We pose the question to ourselves: What are the adjusted series of steps required in order to proceed toward a remediated design strategy for our undertaking? We should be clear that SynBERC has in fact been an adequate venue for conducting first-wave diagnostic work in human practices. We have been free to ask the question of whether or not synthetic biology poses new problems, and whether current disciplinary forms and infrastructures are sufficient. SynBERC, however, has not proven sufficient as a venue for developing strategic responses to these first-wave questions. It thus has not proven sufficient as a venue for flourishing.

7 The Second Wave of Synthetic Biology 2009

From Parts to Ontological Domains

In this chapter we schematically present an account of what in 2009 was designated as the second wave of synthetic biology. We provide this account so as to chronicle significant thresholds in the development of this discipline. On another register, this account occasioned reflection on our part as to the status of human practices. Is there, we wondered, a second wave of human practices?

First Wave: Figuration

From the outset, one of the genre characteristics of synthetic biology has been framing. Such framing produces a figuration that facilitates orienting as well as the construction of overarching narratives of the field's constitution and promised future horizon. In the first wave, such figuration ran ahead of actual experimental practices and scientific results for engineering as well as human practices. In the first wave, the engineering challenges in synthetic biology were typically figured as the opening for the final attainment of the long-promised utility and capacity of genomics and molecular biology. This engineering framing enabled a vision of how basic biological processes might be successfully black-boxed through the use of abstraction hierarchies (i.e., parts, devices, chassis). Such abstraction hierarchies, first-wave proponents held, would allow engineers to ignore, and thereby work around, the complexity and context-dependence of cellular processes. Such a work-around, the proponents further claim, might facilitate a quantum leap beyond the still-elementary capacity to intervene in genomic processes.

In their article "The Second Wave of Synthetic Biology: From Modules to Systems," Purnick and Weiss present what they take to be the utility as well as the limitations of the first-wave approach. The first wave of synthetic biology, they observe, drew on analogies from the development of integrated circuits in electronic engineering as a means of establishing its research priorities as well as its principles for design and construction.[1] As Purnick and Weiss's review demonstrates, these analogies have proven fruitful as an orientation for first-generation work. Sustained efforts to achieve context-independence in the design of biological components as well as emphasis on the goals of standardization and modularization have produced a range of "genetic elements used as components of synthetic regulatory networks."[2] These elements have introduced refined capacities for inflecting and directing genetic expression—transcriptional, translational, and post-translational control. We would add to Purnick and Weiss's review of the technical achievements of first-wave synthetic biology the crucial fact that the analogies to electronic engineering and the discursive framing that went along with those analogies have also facilitated the initial design and organization of synthetic biology as a large-scale project.

First-Wave Limitations

The major challenge today, according to Purnick and Weiss, consists in integrating first-wave parts and modules into "systems-level circuitry." We observe that they continue using terminology from an analogy to electronic engineering. The genetic circuits produced to date are "usually aimed at controlling isolated cellular functions" and not designed for successful systems level functionality, per se. Such a state of affairs might simply be taken as the natural progression of things from the construction of simple well-controlled circuits toward increasingly complex designs and functions.

Purnick and Weiss's diagnosis, however, is that first-wave synthetic biology is blocked by fundamental limits in its design principles and strategies. "Over the past few years," they report, "activity in the field has intensified, as reflected by an increased number of published experimental circuits. Surprisingly, the actual complexity of synthetic biological circuitry over this time period, as measured by the number of regulatory regions, has only increased slightly." The conclusion they draw from this is that "existing engineering design principles are too simplistic to efficiently create complex biological systems."[3] To put it differently, and to put it in our terms, Purnick and Weiss argue that the challenge for second-wave synthetic biology is not only to make bioengineering more *engineerable* but also to make it more *biological*.

The architects and proponents of first-wave synthetic biology could well accept Purnick and Weiss's recapitulation of their design principles and project priorities, and might accept our description of first-wave figuration. One study cited by Purnick and Weiss illustrates this point. Barry Canton,

Anna Labno, and Drew Endy, writing in *Nature Biotechnology*, repeat a standard first-wave premise: "The ability to quickly and reliably engineer many-component systems from libraries of standard interchangeable parts is one hallmark of modern technologies."[4] Canton et al. recognize that "the apparent complexity of living systems" may not "permit biological engineers to develop similar capabilities"; the "emergence" of unanticipated functions resulting from the combination of multiple components may prove intractable to this approach. They stress, however—exemplifying first-wave figuration—that such a possibility should be dealt with through intensified efforts at conceptual black-boxing: "Careful characterization and analysis of such emergent behaviors is needed to support the development of design rules that prevent interactions between devices other than through the defined device inputs and outputs."

If proponents of first-wave synthetic biology would likely accept Purnick and Weiss's recapitulation, they would also likely reject the conclusions drawn. The limited increase in the actual complexity of synthetic biological functionality, it could be argued, is due less to the inherent limitation of the design principles borrowed from other domains of engineering, and more to inconsistent efforts to apply those principles in a standardized and rigorous manner. Such a position would reject the further proposition that new principles are needed that conform more directly to the biological character of living systems. The question for first-wave synthetic biologists, after all, as Evelyn Fox Keller has shown, is not whether or not biological systems actually work like other kinds of engineered systems, but, rather, whether or not biological systems can be *made to work* like other engineered systems.[5] Hence, to repeat the point made above, first-wave synthetic biologists make the strategic decision not to account for complexity and context-dependence in the design of their biological components. Rather, first-wave practitioners emphasize the need to produce techniques and standards that favor—and thereby anticipate and control for—the design and construction of components that maintain acceptably stable levels of performance across contexts.[6]

Whatever the counterclaims to Purnick and Weiss's assessment of the state-of-play in first-wave synthetic biology, it is nevertheless plausible that the use of analogies from electrical engineering and their integration into a figure have so far proved to be limited in their applicability when taken up as a literal guide to organizing biological and industrial work. Such limitations are reflected in the fact that actual research programs in synthetic biology, although benefiting indirectly from first-wave figurations at the level of funding and communication, are actually constituted as projects concentrating on a different scale and type of object. As we have described in chapter 4, and as Purnick and Weiss indirectly demonstrate in their review, work today is proceeding less as a matter of integrating parts, devices, and chassis, and more as a diverse ensemble of projects concerned with parts, pathways, genomes, and systems.

Purnick and Weiss underscore that the accomplishments in the first wave of synthetic biology were nontrivial; significant advances were made in producing small-scale modules designed to inflect processes of translation, transcription, and so on. Nonetheless, there is a real danger that the heuristics, analogies, and design principles that animated the first wave of synthetic biology will stand in the way of developing "effective strategies for assembling devices and modules into intricate, customizable larger-scale systems."[7] In this light (to repeat a quote), Purnick and Weiss argue that "it is possible that existing engineering design principles are too simplistic to efficiently create complex biological systems and have so far limited our ability to exploit the full potential of this field."[8] One of the major pitfalls of the literal use of nonbiological analogies for constituting research initiatives in biology is in mistaking the strategic ambition to black-box biological complexity with the effective capacity to do so. All of which suggests, paradoxically, that first-wave synthetic biology is not yet fully synthetic biology.

Second Wave: Figuration

Purnick and Weiss advance the need, task, and challenge of formulating designs and strategic aims for a second wave of synthetic biology. They underline that this does not mean abandoning the work and insights conducted within the first-wave frame. However, synthetic biology, they argue, should focus on "wholesale changes to existing cellular architectures and the construction of elaborate systems from the ground up."[9] We note that this proposal means that the essential objects of concern for work in such a second wave would be neither parts nor devices nor chassis, per se, although attention to the work done on such units cannot be overlooked.

In order to advance toward a more adequate synthetic biology worthy of the name, Purnick and Weiss proceed by foregrounding, rather than blackboxing, biological complexity and context-dependence. Their strategy is to identify domains within the cell or cellular populations in which biological complexity holds the promise of being manageable and potentially open to strategic leveraging. It is worth noting that these domains, according to Purnick and Weiss, can be identified on the basis of first-wave research. For example, experiments with producing synthetic protein scaffolds in yeast might serve as a synthetically produced cellular domain within and through which novel proteins might be designed and produced.[10]

First-wave practitioners have made the decision to strategically limit attention to these domains within the cell, treating them either as background to, or downstream results of, work on modular components as the crucial objects of interest. Purnick and Weiss's proposal, by contrast, is that strategically such domains can be privileged as the sites and the scale at which problems of biological functioning worthy of investigation and intervention are located. In this light, they argue that second-wave synthetic biology must meet the challenge of shifting orientation, as well as design parameters, so as to confront a more salient range of problems. We argue that such reorientation will require a shift in figuration as well, as we will show in the next chapter.

Ontological Domains

Second-wave foregrounding of complexity and context-dependence introduces a promising technical, conceptual, and design threshold for synthetic biology: the challenge of identifying and targeting the regulation of *ontological domains*. Purnick and Weiss proceed in their proposal by identifying three ontological domains currently being privileged and worked on by synthetic biologists. These domains, they argue, hold the promise for the advance of second-wave synthetic biology. Their method—the way in which they identify and establish connections—can be cast as follows. First, they identify a *design challenge* for synthetic biology. They then pick out an ontological *domain* within which that design challenge can be addressed. They then characterize a core *problem* for which synthetic biology may be well-suited to provide advances in modulation, regulation, or control of such domains. Finally, they specify projects as cases in which these problems are currently being explored through experiments. By connecting design challenges, ontological domains, engineering problems, and research projects, they can provide some tested design parameters that should be instructive for other research projects formulated to investigate these same ontological domains, even if these other projects are concerned with different aspects of the core problem. Purnick and Weiss open up the possibility of articulating candidate design parameters for second-wave work.

Domain 1: Synthetic Ecosystems

A first challenge and major technical threshold for synthetic biology consists in designing multicellular systems whose behavior can be regulated and fine-tuned at the level of the population. First-wave work is often conceived of individual functions and cells in isolation. Purnick and Weiss underscore the point that even within the controlled setting of laboratory practice, cellular organisms are partially determined and directed by interactions with their environment. Such interactions trigger coordinated behaviors on the part of some cellular populations. Identifying and engineering the mechanisms that drive cells' capacity to sense their environments, and connecting these mechanisms to domains that regulate cell-to-cell communication within these environmental milieus, present a core problem and opportunity for second-wave synthetic biology. The problem, in other words, is how to design signaling mechanisms for regulating synthetic ecosystems.[11]

Purnick and Weiss specify several projects as cases of work in this domain, including the present and past work of the Weiss Lab. Among the latter are projects to design and construct artificial signaling pathways in multiple organisms, including AHL, a bacterial molecule that enables the coordination of group-based behavior based on population density, acetate-based signaling in plant bacteria, hormone signaling in yeast, and nitric oxide signaling pathways in mammalian cells. This prior work in leveraging the mechanisms responsible for multicellular communication through the construction of artificial signaling pathways has facilitated the Weiss Lab's efforts to "devise effective design principles for building unique systems with new capabilities."[12] It is based on such proposed design principles that parts and modules composed in a first-wave mode might eventually be refined and integrated into "even more complex multicellular systems with practical purposes."[13]

Domain 2: Application-Oriented Systems

A second design challenge identified by Purnick and Weiss, which follows from the first, concerns the core technical ambition and justification of much of the work being organized under the label *synthetic biology*: programming cells to produce macro-functions that can be used as solutions to current unsolved problems in bioengineering. Such unsolved, application-driven problems include such things as the design of a suite of new methods for systems-level control of gene expression, creation of novel enzymes "tailored to new tasks," and strategies for adapting constructs to distinctive features of host organisms.[14]

Purnick and Weiss highlight two projects as cases of work concentrating on cell reprogramming. First is work being directed by SynBERC PI Chris Anderson. Anderson's project, one of the original SynBERC test-beds, consists of an attempt to design bacteria capable of seeking out and destroying cancer tumors *in vivo*. The project is characterized by Purnick and Weiss as a "living computational therapeutic tool." It harnesses the ability of the bacterium to sense multiple environmental conditions at the same time by using designed "logic gates" to trigger capacities for finding, invading, and killing the tumor cells.

A second case of an application-oriented system is the Keasling Lab's and now Amyris work designing and building synthetic isoprenoid pathways for the production of artemisinin, a precursor molecule for the treatment of malaria. As we described in chapter 5, this same pathway is being reprogrammed to produce a biofuel precursor molecule. Purnick and Weiss underscore that a key technical achievement in the Keasling Lab's work was the ability to get the same pathway to function in multiple host contexts, an achievement that required incorporating specific context-dependencies as design variables.[15]

Domain 3: Minimal Genomes and Synthetic Life

A third second-wave design challenge concerns efforts to construct new synthetic genomes and thereby produce the minimal genomic conditions needed to successfully carry out normal cell function, a challenge often cast as the attempt to create "synthetic life."[16] One engineering problem to which such efforts are oriented is the stated need for a genomic "chassis" of sufficient simplicity to use as a basis and context for the expression of other engineered constructs. The hope and expectation is that a simplified or minimal genome (a genome that has been stripped of those elements that do not seem to be necessary for sustaining cellular life) "will enable synthetic biologists to build less encumbered pathways, resulting in fewer undesired interactions with endogenous systems." Whether or not such a simplified genome might also reduce the ability to leverage certain complex functionalities remains to be seen.[17]

Purnick and Weiss pick out projects as cases that demonstrate the ability to create a programmable synthetic cell by distinguishing divergent design strategies. "The creation of a minimized cell," they write, "can be accomplished using either a top-down elimination approach or a bottom-up forward engineering approach." The first approach is exemplified by the work of biologists at the J. Craig Venter Institute (JCVI). This approach begins with living cells and attempts to determine how much of the genome can be eliminated while preserving cellular viability. Using such an approach for work on the bacterium *Mycoplasma mycoides*, JCVI successfully sequenced, identified gene sequences essential to cell vitality, redesigned or removed non-essential sequences, synthesized, and transplanted the genome of the bacterium. Crucial to JCVI's demonstration of success was that they transplanted their synthesized genome into a different strain of bacteria. Within a small number of population doublings, the synthesized genome had transformed the host cell into an organism closely resembling the bacterium from which the initial genomic sequence was derived. Whether using a top-down approach such as the Venter Institute or a bottom-up approach in which engineers "attempt to create a cell de novo by constructing a membrane-bound compartment and then adding components," an approach taken by several research groups, a goal and hoped-for outcome is to eventually provide synthetic biologists with a "flexible toolbox of minimal genomes" to use as host organisms for their constructs.[18]

Strategic Design Parameters

Having identified privileged ontological domains, Purnick and Weiss are in a position to proceed to specify with more precision the next level of experimental challenges and design parameters for synthetic biology. They begin by naming five "open questions" that to date have been advanced by the critics of synthetic biology as fundamental blockage points. These open questions call for a kind of strategic reversal by way of which the question can be posed as to whether and to what extent blockages, and their underlying conditions, might be reconceptualized and recast as design opportunities. The question is not, for example, whether "cellular noise" renders bottom-up engineering impossible, as some critics have suggested. Rather, the question is this: Can the dynamics of cellular noise be understood in such a way that its functional characteristics can be leveraged in the design process itself? Proceeding in this way, Purnick and Weiss formulate the next stage of work in synthetic biology as a challenge of converting ontological obstacles into *strategic parameters* for second-wave *design*. We present three of these "open questions" here to exemplify Purnick and Weiss's strategy.

Parameter 1: Characterization, Standardization, Modularization

First-wave synthetic biology has been characterized by a central premise and central question. The central premise is that, following common practice in other engineering disciplines, biologists need standardized modular components that are "easily combined to form larger systems" across multiple contexts. The central question, then, is this: How can synthetic biologists combine many basic components effectively so that cellular behavior is optimized and "made to order"?

A first and widely discussed response to this premise and this question has been the elaboration of the BioBrick standard for the design and physical assembly of genetic elements.[19] As Purnick and Weiss note, and as has emphasized by a number of other researchers, extending the BioBrick standard beyond physical assembly to support predictable and reliable functional composition remains an ongoing challenge.[20] A frequent critique of the BioBricks approach has been to highlight the fact that characterization, standardization, and modularity are significantly affected by cellular context. This critique argues that in order for a part to be used in a fashion similar to the interoperable components characteristic of other engineering practices, biologists and engineers will need to know how the behavior of a part changes with regard to genetic, host, and cellular contexts.

Purnick and Weiss recast this critique and blockage point as an opportunity to pose the question of how synthetic biologists might go about the work of designing or refining genetic components in view of the ontological domains within which these components will be assembled and put to work. The point is not to abandon the use of standardized parts or modules, but rather to align standards for design and assembly with functional requirements that are themselves determined to a significant degree by intracellular, intercellular, and extracellular environments. Such environments, in other words, need to be foregrounded during the design process. Such foregrounding, they suggest, would premise the construction of synthetic components on the eventual need to integrate those components in specified "systems and contexts of interest."[21]

Parameter 2: Noise

A second point of critique and blockage in the first wave of synthetic biology has centered on the problem of "genetic noise" in cellular systems, and whether or not, given the amount of noise in these systems, it is "reasonable to expect that we can construct reliable, robust, and predictable systems."[22] Noise—understood as unwanted interference with, or perturbation of, desired signals or interactions—has many sources within the cell. These sources include "extrinsic environmental variations, fluctuations in gene expression, cell cycle variations, differences in the concentrations of metabolites and continuous mutational evolution." Hence the question: Given the range of variables producing noise in a cell, and given the fact that even at the level of the function of a single gene such noise can interfere with the "stochastic nature of biochemical reactions," will it be possible to manage and regulate intra- and intercellular systems in second-wave synthetic biology, as Purnick and Weiss propose?[23]

Purnick and Weiss stress the point that noise in genetic systems is usually assumed to negatively impact synthetically designed cellular processes, particularly processes that require a high degree of specificity and fine-tuning. As such, previous attempts to deal with noise in synthetic biology have centered on designing strategies whereby noise can either be blocked or at least significantly attenuated. The Weiss Lab's work on the production of synthetic signaling pathways, however, has shown that cellular noise, far from been an extrinsic and interfering factor in successful cell signaling, actually "has an integral role in maintaining stable cell densities." The challenge, in this light, is to determine under which conditions and to what extent cellular noise actually needs to be leveraged in order to achieve a desired level of functional control in engineered systems. Further study of noise in relation to regulatory networks, signaling oscillations, feedback systems, and so on would likely prove fruitful in determining how a population of cells can successfully "sense and respond to their environment . . . during times of stress."[24]

Parameter 3: Epigenetics

A third blockage point for first-wave synthetic biology concerns epigenetics. Purnick and Weiss define epigenetics as "the heritable changes that propagate without changing the underlying DNA sequences." Similar to the other parameters for second-wave work that they identify, epigenetic effects may be deleterious for the controlled expression of designed genetic circuits, and therefore some first-wave practitioners have devoted considerable efforts to conceptually and materially black-boxing them.[25]

Purnick and Weiss propose that epigenetic properties might be leveraged in order to design cells "that can synthetically maintain or modulate epigenetic 'memory' on demand." The advantage of such designed cells is that they would be able to transfer properties between cell generations through genetic as well as non-genetic mechanisms. This means that properties such as gene states, which effect gene expression, could be used to "propagate information from a cell to daughter cells." They ask: "if an organism has already propagated an epigenetic modification, how can synthetic biologists change or release the modification and return a gene to its default state?" Epigenetics offers one possible answer. In short, the complications of epigenetics, which encouraged black-boxing in first-wave synthetic biology, are recast as possible design elements for the engineering tool kit.[26]

An initial figural lesson from second-wave synthetic biology can be cast as follows: Everything does not need to be designed in order to achieve total control. In fact, in living systems today, it is not possible to achieve such total control. Given Purnick and Weiss's program, this fact is not an insuperable obstacle to advance in biological engineering, but instead points the way toward making engineering biological. After all, evolution itself can be approached as providing a long and wide-ranging set of experiments, results, and lessons about the variability and function of living systems. More specifically, this orientation directs attention to the parameters according to which living systems themselves have achieved functional specificity.[27]

Conclusion: From Things to Problems

Echoing Max Weber, we underline that the proposed shift in strategy articulated by Purnick and Weiss represents a movement from an attempt to harness "the 'actual' interconnection of 'things'" (such as promoters, ribosome binding sites, and switches) to a more focused attention to "the conceptual interconnection of problems." Examples of the latter would include the following: How does the cell regulate complexity and timing of protein interactions? What does functional generalizability mean in a biological environment? Or how do signaling pathways determine the activity of cell populations?

In this light, our claim that the initial figuration of synthetic biology is not yet fully synthetic biology is less paradoxical. After all, again following Weber, a "new 'science' emerges" only where "new problems are pursued by new methods and truths are thereby discovered which open up significant new points of view."[28] We argue, following and reinterpreting Purnick and Weiss, that a shift from attention to first-order *things* to second-order *problems* and *concepts* stands a better chance of opening up new perspectives for synthetic biology as well as human practices. Our informed wager has been that such perspectives should facilitate the introduction of refined design and experimental parameters for the next wave of our work.

It is tempting to use Purnick and Weiss's categories in a literal fashion for exploring the possibility of a second wave of human practices. After all, as we have shown, their framing of second-wave synthetic biology has much to offer in terms of opening up new scientific problems and approaches. We will avoid this short-cut, however, for two reasons. First, in contrast to Purnick and Weiss, who provide cases of second-wave work already under way, we emphasize again that a second wave of human practices has not yet found a venue that facilitates systematic experimentation. Although most of the main actors in human practices have proposed the need for moving beyond the first-wave diagnostic work summarized in the previous chapter, these proposals have not yet succeeded in settling on objects and concepts, or strategies. More important, even though aspects of these proposals may eventually form part of a second wave of human practices, the extent to which these strategies will open up new methods, truths, or perspectives is far from clear.

Second, and more important still, our first-wave diagnostic work has indicated that direct analogies per se need to be adopted with great caution. They need to be treated skeptically, that is to say, following the term's etymology, inquired into before any stabilization or acceptance is decided upon. Thus, returning to our first point, it is through reflection on actual research taking place in human practices that our design parameters for second-wave work can be established.

8 A Mode 3 Experiment

Figuring Dual-Use: From Safety to Malice

> To grasp a situation solely in moral terms is therefore ulti-
> mately immoral: it implies a certain laziness, a refusal of the
> effort which is necessary in order to discern in the situation
> those specific features which one nevertheless needs to know
> and understand in order to act better.
>
> FRANÇOIS FLAHAULT[1]

Our diagnosis is that in the arena of the biosciences and the human sciences in which we have been working, there are two different and heterogeneous figures currently available: biopower and dual-use. The concept of a *figure* designates a form of connections among events, actors, discourses, practices, and objects so that a more or less stable and integrated ensemble is produced. The form of this ensemble is such that the significance and functions of the ensemble cannot be reduced to its constitutive elements. Figuration thus also designates a way of connecting elements into an ensemble where the significance and functions of each element depends on, though may not be reducible to, the form produced by the connections. Figuration involves a kind of synthesis—the production of a composite whole whose logic of composition cannot be reduced to its constitutive elements. If figuration designates a way of connecting and synthesizing elements, the resulting ensemble can be designated as a *figure*.

The terms *figuration* and *figure* have a long history extending back to the Greeks.[2] In our present work, we find pertinent and helpful Erich Auerbach's concept of figural interpretation.[3] Figural interpretation, as Auerbach describes it, is a method of taking up reality in which connections are established between "two events or persons in such a way that the first signifies not only itself, but also the second, while the second involves or fulfills the first." For Auerbach, the poles of the figure are integrated in and by a shared temporality.[4]

Dual-Use: Security and Malice

One of the central conundrums of the contemporary biosecurity landscape is what to make—semantically, rhetorically, politically, and strategically—of the figure of dual-use? Although there is doubtless a prehistory to the term *dual-use*, since September 11, 2001, though the word remains the same, the term's concept and the referent have changed. Those changes have been assembled into a coherent figure with semantic, rhetorical, political, and strategic ramifications. This work has been achieved through a specific type of semantic practice in which disparate events are connected and brought into prominence through a *figuration*. To put this claim another way: What was once a trope (indiscriminately an analogy, a metaphor, a simile) has become a figure. Ramifying out from that act of figuration, and in seeming conformity with it, is a whole series of strategic developments.

In this case, the figuration turns on a core distinction—that there are *good* and *bad* uses of science and technology. This distinction has rapidly been taken for granted as both self-evident and the key to the nature of biosecurity today. Once such self-evidence has been established (through figuration), it is logical and even imperative to mobilize resources to ensure that the good can be distinguished and segregated from the bad and thereby contained. By making these connections, the figuration is given form materially, organizationally, and infrastructurally.

The reasons for the ready acceptance of the dual-use framing of biosecurity are various. However, none of the reasons, upon reflection, are obvious. None of the reasons, upon even closer inspection, are compelling. Consequently, it is worth the analytic effort to disaggregate some of the diverse events, practices, problems, and capacities that have been bundled together. By so doing, we are in a position to better understand the specific relations of knowledge production, power relations, and affect at issue today in the field of biosecurity, especially as it interfaces with the biosciences.

The biosciences (and others as well) are currently conceived, designed, carried out, financed, and—most importantly—justified with the goal of intervention into natural processes. Further, the mode of construction of scientific objects in the biosciences is not only constructivist (all sciences are constructivist in a noncontroversial sense) but guided by a small set of explicitly instrumental norms and goals. Some fear those interventions, others applaud them, but both sides often unreflectively concur that what the biosciences are doing is and should be fundamentally instrumental, guided by a norm of intervention. The opponents of "genetically modified organisms" almost never object to environmental cleanup.

Currently, this interventionist and instrumentalist norm is embedded in claims to the importance of health, environment, and security. These claims fall neatly into the category of the biopolitical. The biopolitical, following in the line of Michel Foucault, is that strategy whereby fundamental biological processes enter into and are reconfigured by projects of social control and improvement.[5] The *biopolitical* is a general term that groups specific historical attempts to bring diverse discordant phenomena into a common frame. For example, famine, urban riots, and grain shortages over a long term became a topic of biopolitical study and state intervention. Today controlling the spread of biotechnological capacity and channeling it into disease control and environmental amelioration similarly require new forms of knowledge, new technocrats, and new organizations.

Representing and intervening, knowing and doing, discovery and invention today are taken up under the sign of intervention.[6] Although this may seem obvious and beneficial, it actually entails far-reaching ontological, ethical, and political ramifications. We are obliged to recognize and reflect on the ramifications of this now dominant form of understanding and practice. Examining the semantic mangle analytically offers the solace of thinking more clearly about where we find ourselves today.

Mode: Communication

One of the characteristics of the decade or so of genome mapping was its proliferation of discourses of hope and despair about the promises and dangers of the emerging genetic technology. This discourse was deployed as a funding device by those like James Watson or Francis Collins, who had the responsibility of selling and maintaining the large program to the U.S. Congress.[7] The small set of tropes spread far beyond such directly tactical situations, however, and were taken up by the mass and scientific media alike with surprisingly little nuance. The majority of the critics of the project, while reversing the valence of their claims basically (although there are exceptions), were operating within the same tropological and semantic field. The tropes of the "book of life," the "holy grail," "Frankenfood," or "back door to eugenics" all seem rather quaint today.[8] They seem quaint because in retrospect, looking back on the recent past, they had so little purchase—at the scale they had promised—on what actually happened in labs or clinics or farms. Although that claim is contestable, the fact remains that versions of these redemptive or apocalyptic tropes continue to circulate. Tropes, it seems, are not refuted but only replaced. Or they fade into the communicative pollution of accepted truths while refusing to engage in any venue in which their claims might receive warrant or rejection.

New and improved technologies as well as new knowledge of molecular and cellular biology introduced new problems and required new research as well as new technologies and venues to facilitate that research. Inquiry is like that. The sequencing of the human and other genomes at the turn of the twenty-first century underscored the reality of the massively increased capacities for genome sequencing. The genome-sequencing projects were conceived as fundamentally technology projects: could the cost, accuracy, and efficiency be achieved that would make the mechanical task of sequencing hundreds of millions of base pairs economically feasible? A decade's work demonstrated that the answer was affirmative. And that trajectory has continued with lowered costs, higher accuracy, and many more dispersed sites of activity, larger databases, and a fledgling synthesis industry.

Although the project was cast as a techno project (in line with the American sublime), a great deal was learned about genomes. One of the most astounding discoveries was the extensive homologies in the molecular constitution of living beings.[9] For example, the large number of oncogenes shared by *Drosophila* and *Homo sapiens* points to a depth of shared substance and function between otherwise vastly different organisms that is breathtaking. Further, the unexpected and unanticipated discovery of the small number of genes in the human genome indicated that other mechanisms of control, regulation, and production beyond the quantity of the genes had been developed evolutionarily. Although all molecular biologists now subscribe to this claim as if it were self-evident and not highly significant, this underestimates the scale and importance of biological understanding that the sequencing projects provided. For example, the discovery of such mechanisms as interference RNA—previously unknown even though it is a crucial component of how living beings regulate molecular function—was rapid. This discovery opened up, among other things, a range of new potential modes and sites for functional intervention. A sector of the biotechnology industry and its venture capital base rapidly developed to exploit this discovery. Although these modes and sites were immediately coded as potentially therapeutic, upon reflection it would be hard to maintain that they could be limited to hypothetical cures for cancer or environmental cleanup.

Metric: Security

Although the 9/11 attacks on the United States can be taken to be the efficient cause for unleashing forces that have been strategically used so as to transform the previous Cold War and disaster-preparedness security environment, it is relevant that these attacks had nothing directly to do with capacities of biology or biologists. However, the attacks did open up discursive and strategic spaces. One of the primary responses to this event and to this opening were attempts to articulate a new floating signifier of the Other and the Enemy to substitute for those of the Cold War. Diverse actors and affects entered into a semantic and political field suddenly opened for intervention and framing.

Subsequently, the debate during 2001–3 about the existence of weapons of mass destruction in Iraq provided a specific connection (fictive or not) to the ambient political rhetoric of fear and security, thereby connecting geopolitical issues to biological and chemical (as well as nuclear) capacities. The mendacity of the way the particular charges of WMDs were deployed and the uses to which they were put, however, masked or overrode for many the fact that the existence of chemical and biological weapons was, or had been and plausibly could be, a reality. Saddam Hussein had used chemical weapons against his enemies (and their population base) in Iraq and Iran. Further, the dismantling of the Soviet Union had revealed its massive bio-weapons program as well as the existence of large numbers of biologists (thousands) with bioweapons experience who during the 1990s had become part of scientific diasporas.[10] Thus, by 2003 the technical capacity to make bioweapons was a reality. Significantly, a corps of scientists had demonstrated that they were willing to undertake scientific and technological work on bioweapons. Further, it must be said that we do not know with clarity what kind of research and development of bioweapons or biodefense capabilities were being carried out in sites such as Fort Detrick in the United States and presumably elsewhere (in the UK, for example). Although we do not know all the details, it is clear that facilities and resources for such research and development did exist. They have been expanded dramatically after 9/11. In sum, the Soviet and American programs (among others) were not dual-use—they were instrumentally proliferative.

Parameter: Malice

Analytically the argument about dual-use has two sides: an objective side and a subjective side. In reality, of course, these two are not partitioned off; they form a tangled and confused web. On the subject side, the argument confronts the conceit that the danger of nefarious use of contemporary scientific capacities can be cantoned off into the "bad guys" (terrorists or psychopaths or some combination of the two) and the "good guys" (usually boys and girls wearing lab coats, government officials wearing suits and ties, or venture capital money people wearing either jeans and black T-shirts or expensive suits), who, it by now goes without saying, only want to do the right thing. To suggest otherwise is bad manners; to question the distinction occasions indifference—a turning of heads to see if anyone else in the room has something helpful to contribute—or belligerence—hostility toward those who are wasting important time and complicating the issue for no known end.

We take the concept of "malice" from the title from a small book, *La méchanceté*, by a psychoanalytically oriented French literary critic, François Flahault. The book treats Mary Shelley and Frankenstein and what Flahault convincingly considers to be a genealogy of misreadings of this by now primordial text of our security concerns.[11] Frankenstein, however, is not at issue here. Although "malice" is an accurate translation of the French *la méchanceté*, it does not quite capture the phenomenon that Flahault explores. His book is an attempt to identify and analyze the fact that good intentions, or what are claimed to be good intentions, in certain situations and under certain conditions, can operate in a mode that is nefarious. Essentially it is nefarious because it rules out complexity and fundamentally therefore rules out reality. While such simplifications can be helpful, even necessary, in navigating (or producing) the shoals of the world's ethical topography, when turned into the metric of a doctrine, it would be comic if its ramifications were not so pathetic and, at times, even tragic.

The proclamation of good intentions, the drawing of clear and distinct lines between good and bad, itself requires a much more unsentimental examination and cannot, as the expression goes, be taken at face value. So, what are the wellsprings of malice? Flahault begins his analysis by underlining the significance of even asking the question in the first place.

The question is important because not only does it point to self-illusion but equally supports belief in illusions widely shared. These illusions are catastrophic, as has often been said, whenever they guide the political ideal of entire societies.[12]

Flahault is referring to a vast literature on the political fortunes of various regimes of the right and the left in the twentieth century whose self-proclaimed and militant good intentions directed at the construction of a new and better society and state, peopled by new and improved men and women, feel obliged to rid society of dangerous individuals and groups. Furthermore, and here we are perhaps closer to the present world with its NGOs and humanitarian organizations as well as its fundamentalisms currently taking religious form around the globe, when the world is divided into friends/foes, or suffering versus politics, danger lies ahead. In situations of danger, good intentions will frequently establish a structural blindness to possible ramifications of actions or structures. By so doing, the well-intentioned of the world thereby obfuscate the need to transform dangers into risks if one is to act with foresight and maturity.

Flahault argues, and we are in full concord with the direction of his argument, that such regimes of piety and danger are also dangerous when they collapse, since it is then that the well-meaning, who had learned to associate the desire for good with idealization, either lose the impulse which animated them or else, in order not to lose it, preserve within themselves an idealization which they only pretend to renounce.[13]

Flahault offers a pocket history of malice by contrasting two paradigmatic figures in which evil is given form. On the one side is Saint Augustine, whose doctrine of original sin placed malice within each human but offered a hope of redemption. On the other side is the figure of the Enlightenment characterized as a reaction to Augustine and his version of Christianity, counterposing an affirmation of the inherent goodness of humanity requiring improvement of the self and of society as the path to a this-worldly redemption. By externalizing malice, a vast potential for work in this world—of a certain type—is mobilized. By externalizing malice, questioning the inner wellsprings of malice is relegated to other traditions. Once malice is externalized—there are good and bad uses of technology—once reform and the hope for progress falter, no other resources are at hand for understanding or even addressing the issue.[14]

This externalization and self-idealization lead not only to a certain smugness and an associated complacency. Such a position holds that the malice is elsewhere; we can train scientists to recognize it and denounce it, in good conscience. Flahault characterizes a likely ramification of this externalization as "Puritanism." This externalization leads to outreach programs, to vulgarization efforts, to safety drills, and to vague pronouncements about dual-use.

Such tropes and their associated practices actually work against self-improvement. Since it is only the Other who needs policing or improvement, there is little time available for self-improvement in domains other than one's subspecialties and one's leisure activities. This focus is dangerous since "self-improvement involves a process of de-idealization."[15] And de-idealization is the first step toward pragmatism. Such pragmatism entails testing, evaluating, and understanding the limits of one's findings as well as their strengths. In a word, it is a vital component of a scientific ethos.

Venue and Equipment: Commissioning Dual-Use

Although we are living in a period of the massive proliferation of nongovernmental agencies, we also live in a moment in time where an older form of venue still retains a certain distinctive importance: the advisory commission. Advisory commissions are designed to eschew overt political advocacy. This demurral allows them both to claim a moral and/or a knowledge-based high ground; to set forth conditions, usually discursive, for modes of discrimination, to lay out seemingly distinct. Advisory commissions are called into being in principle to present a map of possible directions to adopt for decision-makers without committing anyone to any particular course of action. By limiting their role to the task of laying out a range of options, commissions establish a certain position of authority through their claim to impartiality and balance. That authority, of course, can be denigrated or its recommendations ignored, as we saw with the so-called 9/11 Baker-Hamilton Commission. Today the members of the 9/11 Commission appear with some regularity on media talk shows to remind the audience that their recommendations have been ignored. Such appearances are occasioned when recognizable repetition of a known risk is enacted.

....

The term *venue*, taken in a technical sense, designates a setting in which specialists work on design and synthesis. A venue is not a neutral scene in which specialists work, nor is it only the site within which a given mode of composition is advanced. Rather, it is a facility. Venues may have been already stabilized or institutionalized, they may coincide with the articulation of practices and organizations, or they may emerge through the practice of equipmental composition. That is to say, when composition is successful, the venue facilitates rather than obstructs the design and synthesis of specific interfaces. Consequently, there are venues in which particular interfaces are more likely to be obstructed than facilitated.

....

The significance of advisory commissions as a venue for biosecurity today is that they are framing dual-use organizationally and practically. In fact, to be more precise, what they are actually doing is establishing the figure of dual-use, which can then be useful and have effects when it is subsequently circulated and expanded. Of course, commission members would not recognize themselves as working on figuration. This produces a certain blindness with regard to the ramifications of the advisory work undertaken.

Commissions and reports have played a leading role in supporting the advisory framing of the biosecurity landscape in the early twenty-first century. By 2010 three such commissions, staunchly advisory but with a certain para-governmental standing, have been influential in solidifying the current semantics and rhetoric, culminating in the establishment of the figuration of *dual-use*. This arrangement is a typical American combination of financing from a private foundation, which enables government officials, prestigious scientists, and engaged observers to come together into a common venue. The product of such commissions is a nicely turned-out report. Conveniently, such a report has the standing neither of law nor of policy; rather it functions to establish talking points and thereby to structure the field of communicative rationality in which future actions will be evaluated or claims and counterclaims will count as true or false.[16]

If the overtly political is eschewed, and the enactment of policy recommendations and their implementation often a long way off, the authority of these commissions (as is the case with the NGOs) derives from and contributes to the inscription of the *ethical*. The ethical discourse of commissions, of course, is admonitory, whether urging caution, pressing for virtue, or simply declaiming the need for an agenda that normal political channels cannot or will not provide but that, at least in principle, lies within the mode of jurisdiction of the government or other legitimate corporate bodies. Of course, anyone intent on ignoring these guidelines would be unconstrained to do so; advisory commissions, after all, end up being advisory.

Fink Report: Dual-Use between Biopolitics and State Security

The first in a series of commissions dealing with the biosciences and security was the Fink report on "Dual Use Research."[17] In the preface, Gerald R. Fink (chair), professor of genetics at the Whitehead Institute for Biomedical Research at the Massachusetts Institute of Technology, lays out the commission's basic charge. Fink notes that this report was not the first to deal with the relations of security and science, but it was the first to deal with security and the life sciences. Conditions have changed

> the discovery of nations with clandestine research programs dedicated to the creation of biological weapons, the anthrax attacks of 2001, the rapid pace of progress in biotechnology, and the accessibility of these new technologies via the Internet. The goal of this report is to make recommendations that achieve an appropriate balance between the pursuit of scientific advances to improve human health and welfare and national security. In preparing this report our Committee examined ways by which the spread of technology, methods, materials and information could be limited to constructive activities concerned with medical progress.[18]

The paradigm case that defines all of these commissions is the famous conference at Asilomar in 1975 at which a self-selected group of elite scientists debated and articulated a framework for confronting fears around recombinant DNA. Essential to the Asilomar paradigm was a rhetoric of

reassurance that the scientists were well-intentioned, that they were working to advance science and improve the public good—and, above all, given
these professions of virtue, that the public should have confidence in the
scientific establishment. The keystone to the paradigm was actually symbolic: that all dangers could be essentially reduced to issues of safety. Safety
could be reduced to lab regulations and some limited and temporary oversight, operating without strong legal or criminal sanctions, as to the kinds
of experiments that could be undertaken responsibly. On the level of safety,
the guidelines have proved successful. On the level of discursive containment, they can be trusted as well.

On one side, the Fink Commission takes as a basic supposition of its
work that the political and economic conditions for the practice of the life
sciences are now unlike those that were in force at the time of the Asilomar Conference. It is no longer the case that the biosciences are practiced
uniquely by elite scientists in secure facilities. On another side, however,
the basic logic of Asilomar, turning on strategies of containment, remains
in force. In this light, the problem is not physical containment in order to
prevent against accidents and ensure safety. Rather, the problem is to distinguish between good and bad actors and intentions, and to elaborate possible strategies by which the former can go about their work, while the latter are excluded from the global circulation of knowledge and materials.
Hence, the containment strategy, taken up under the metric of security
and the parameter of malice, remains in place, but its object and equipment
have changed. And, therefore, the wholesale transposition of the Asilomar
figuration, while convenient for extending the privileges and power of the
biological sciences, is blind to the shift from safety to malice. Containing
malice and containing safety are not the same thing.

In fact, the commission focuses on biotechnology and not the biosciences.
Although it advocates "open communication" and the "sharing of ideas," it
makes no mention of the vastly more important presence of patents in both
the industrial and academic world. Asilomar predated the famous Bayh-
Dole Act of 1980, which mandated consideration of commercialization of
federally funded science. Such a horizon of commercialization had become
so much a part of the landscape in the early twenty-first century that it is
not even mentioned. Hence, the further instrumentalization of research is
also taken for granted by the commission. Indeed, rarely do such commissions make space for the pursuit of free science in a free society. Of course,
we have no doubt that at least in principle the members of these commissions are in favor of free science. That being said, they are not created in
order to bring such variables into play.

The rhetoric of containment has proven to be efficacious, at least until the advent of new security concerns following 9/11. As Fink puts it: "But now the nation faces a different problem: the intentional use of biotechnology for destructive purposes." Whereas the Asilomar participants identified the main dangers as unintended consequences, especially the release of genetically modified organisms, the Fink Commission recast the dangers as intentional.

> This task is complicated because almost all biotechnology in service of human health can be subverted for misuse by hostile individuals or nations. The major vehicles of bioterrorism, at least in the near term, are likely to be based on materials and techniques that are available throughout the world and are easily acquired. Most importantly, a critical element of our defense against bioterrorism is the accelerated development of biotechnology to advance our ability to detect and cure disease.[19]

The report's conclusions are characteristic of the genre in their vagueness. In our view, the work they have accomplished is to contribute to the establishment of dual-use as a figure taken up according to the parameter of malice, within a metric of security, and a mode of communication. The work that this and subsequent commissions really accomplished can be summarized as follows:

1. Figuratively established the need for an *affect of vigilance* concerning a heightened attention to intentions.
2. Figuratively identified the need, given these conditions, for an *ethic of responsibility* in which codes of conduct become the order of the day for the well-intentioned insiders and their future colleagues.
3. Figuratively authorized a self-regulatory *mode of jurisdiction*.
4. Figuratively extended a purely technical, instrumental *mode of veridiction*.

NSABB: From Figure to Equipment

The Fink National Research Council (NRC) report made an important contribution to the development of biosecurity policy. Prior to the release of the NRC report, the Bush administration had already initiated a parallel examination of security issues in biological research. This executive branch evaluation was in progress when the report was published. The NRC report recommended a series of actions to improve biosecurity in life science research, one of which was the creation of a standing governmental advisory body. Among other things, one of the things such an advisory body could accomplish would be the transformation of the figure of dual-use into practical equipment. In a nontechnical language, this simply means establishing the practical mechanisms by which the report could be put into practice. In fact, these seemingly trivial mechanisms have long-term ramifications of great importance.

For example, the mechanisms created under the auspices of the Human Genome Sequencing projects—the Ethical, Legal, and Social Implications program, or ELSI—subsequently became the norm for bioethical engagement with the life sciences both in the United States and in Europe. This norm consisted in a downstream positioning. This location, plus broad general principles emphasizing educating the public about the good of genomics research, did its real practical work through support of projects for clinical counseling, the production of documentaries, and the like. We call codified practice *equipment*. Such codification is brought into being in relationship to, and aligned with, a figuration of the life sciences and society, in this instance.

....

Pragmatically, to use our technical language, this demand to bring "science and society" into a common frame is a demand for contemporary equipment. The term is an adequate English translation of the classical term *paraskeue*. Equipment, though conceptual in design and formulation, is pragmatic in use. Defined abstractly, *equipment* is a set of truth claims, affects, and ethical orientations designed and composed into a practice. Equipment, which has historically taken different forms, enables practical responses to changing conditions brought about by specific problems, events, and reconfigurations.

....

In the case of dual-use, a first generation of equipment was established through the work of the NSABB—the National Science Advisory Board for Biosecurity. The NSABB was created as part of the Office of Biotechnology Activities' Dual Use Research Program, whose mission is "the development of policies addressing life sciences research that yield information or technologies with the potential to be misused to threaten public health or other aspects of national security." As it is chartered, the NSABB differs somewhat from the body called for by the Fink report, but its aims and functions are similar to those envisioned by the NRC committee.

The NSABB advises all federal departments and agencies with an interest in life sciences research. The board recommends specific strategies for the efficient and effective oversight of biological research cast as dual-use. Significantly, this includes the development of guidelines for institutional review of dual-use research. The board's mandate is to consider both national security concerns and the needs of the research community when providing guidance and recommendations to the federal government.

The following federal entities were represented in the board's deliberations:

- Executive Office of the President
- Department of Health and Human Services
- Department of Energy
- Department of Homeland Security
- Department of Veterans Affairs
- Department of Defense
- Department of the Interior
- Environmental Protection Agency
- Department of Agriculture
- National Science Foundation
- Department of Justice
- Department of State
- Department of Commerce
- National Aeronautics and Space Administration
- Intelligence Community

Given the establishment of dual-use as a figure, the NSABB began its work by specifying what kinds of objects fall within its purview:

The NSABB has proposed defining "dual use research of concern" as research that, based on current understanding, can be reasonably anticipated to provide knowledge, products, or technologies that could be directly misapplied by others to pose a threat to public health, agriculture, plants, animals, the environment, or materiel.

The NSABB has also proposed a series of experimental outcomes that should be given special consideration for their dual use potential. The public will be invited to comment on this definition, and then the U.S. government will consider whether and how to adopt this definition as formal policy.[20]

In view of this object and their double proposal for how to proceed, the NSABB took as their first order of business (1) determining the extent to which new synthesis technologies intensify the problem of dual-use, and (2) the extent to which current regulatory frameworks "safeguard against the misuse of this science."

In December 2006, in a report entitled "Addressing Biosecurity Concerns Related to the Synthesis of Select Agents," the board published its findings, along with a first series of recommendations. Among other concerns, the report gives particular attention to the difference between "traditional" and "non-traditional" means of engineering DNA, and thereby to the fact that current U.S. government "Select Agents" lists do little to block or even forestall "ease of access" to potentially dangerous genetic materials. In order to produce toxic or pathogenic genetic or genomic sequences by "traditional recombinant DNA technologies," bioengineers need to have access to "naturally occurring agents or naturally occurring nucleic acids" contained within those sequences. Under these circumstances, the U.S. government's select agents list, which provides the genetic sequences for known pathogens, establishes a more or less adequate mechanism by way of which access could be restricted. Providers of such naturally occurring agents or nucleic acids could screen out orders for all but a select range of university, government, and industry labs.

Synthesis, however, allows for the *de novo* production of these same dangerous sequences without access to the naturally occurring materials. Moreover, and equally important, it also allows for small permutations of dangerous sequences to be designed. These permutations mean that researchers can create sequences that are not genetically homologous with items on the government's select agents list while, in principle, being equally dangerous. "The behavior and properties of the expressed product of any synthetic genome that varies at all in its sequence from the exact sequence of a Select Agent 'type strain' may be difficult to predict." In this light, even if synthesis companies were to screen orders for potentially dangerous sequences using the government's list, there could be no assurance that such an exercise would be worthwhile.

In light of this current lack of capacity to identify all potentially dangerous synthetic sequences, the report asks: "What are strategies and mechanisms that might prevent or mitigate potential misuse of synthetic genomics while minimizing restrictions on the beneficial uses of this important field of science?" Among the recommendations in the report, several stand out. First is consciousness-raising. Given the threat of "misuse," investigators and synthesis companies are admonished to know what they "possess, manufacture and/or transfer in order to comply with the SAR [Select Agent Registry]." The effectiveness of such consciousness-raising will depend not only on clarification of what's included in the current registry and the establishment of uniform screening procedures among synthesis providers, but also requires "the USG to fund the development of improved sequence databases and software tools, enhanced understanding of virulence, and improved framework for interpreting sequence screening results." Lastly, the report underscores the need to convene groups of experts from the scientific community (U.S. and non-U.S.) to conduct an "open and in-depth examination of the Select Agent classification system" and to develop an "alternative framework."

In short, the figure of dual-use seems to require that the biotechnical state of play (traditional vs. nontraditional engineering) be aligned with current standards for identifying dangerous things. Where that alignment proves insufficient, scientific expertise must design procedures whereby a better alignment can be achieved. Although "restricting access to new sequence information about Select Agents" may not prove to be feasible, experts may yet be able to provide screening procedures to deny access to the agents themselves.

Equipmental Platforms: Anchoring the Figure of Dual-Use

The Alfred P. Sloan Foundation funded and catalyzed the above commissions and reports (Fink, NSABB). Additionally, they supported a series of extensive workshops that focused in on more technical detail than either of the other commissions. The challenge, in our terms, was how to transform the recommendations of expert advisers into platforms for technically sophisticated and operational responses to dual-use. These meetings and their recommendations were coordinated in partnership with experts from the J. Craig Venter Institute, MIT, and the Center for Strategic and International Studies. The presence of major stakeholders at these meetings was an indicator that the ground-clearing preliminaries had been accomplished and that recommendations could possibly be implemented.

One outcome of these meetings was the publication of a report entitled "Synthetic Genomics: Options for Governance."[21] The stated goal was "to formulate governance options that attempt to minimize safety and security risks from the use of synthetic genomics while also allowing its development as a technology with great potential for social benefit." Thus, the report remains fully within the figure of dual-use. Though the report gives careful attention to technical details, that care and attention serves to concretize and anchor the figure of dual-use. The significance of the report and the principal outcome of the meetings were the generation of a more technical determination of the figure of dual-use as a problem and the discrimination and organization of appropriate standards and activities.

Safety, Security, Preparedness

The report on governance identifies three kinds of security challenges associated with synthetic genomics. In our view, the report's strength is this recognition of diverse problems, albeit grouped within a dual-use figuration. First, it points to the expansion of dangers and risks connected to the intensification of synthesis technologies. The report frames these trends as technical issues of safety. Second, it gestures to changes associated with contemporary political environments, particularly potential malicious users and uses as well as the increased access to know-how through the Internet. Third, the report recognizes that there is a horizon of emergent challenges, which by definition cannot be fully known in advance.

The primary challenge for those working on the governance report concerned intensification of existing dangers arising from the increased capacity of DNA synthesis. Recent innovations in synthesis technology vastly expand the capacity to produce ever-larger specified sequences of DNA more rapidly, at lower cost, and with greater accuracy. Previously, these trends have been framed as issues of safety, which can be addressed through technical solutions. The report carries that framing forward, reinterpreting it in a dual-use framework.

Though focusing on the increase of technical capacities and the possibility of correlated technical containment, the report also gives attention to a new range of potentially malicious actors and actions (i.e., terrorists/terrorism) that must now be taken into account by those seeking to govern scientific domains, and the Internet and other new media that provide global access to technological know-how and scientific knowledge. Such malicious actors and access cannot be addressed using existing models of nation-specific regulation. The report's diagnosis equivocates as to whether new political milieus merely intensify existing challenges, or whether they produce qualitatively new problems that would require qualitatively new solutions. Either way, these challenges cannot be adequately dealt with through an existing frame of safety; but require a shift to a security framework.

The report briefly alludes to the issue of complex uncertainty. By definition, all scientific research is characterized by a measure of uncertainty with regard to whether its experiments will succeed. While some safety and security dangers are presently identifiable, we lack adequate frameworks for confronting a range of new dangers that fall outside of previous categories. Such frameworks would need to be characterized by vigilant observation, forward thinking, and adaptation. Challenges related to uncertainty should be framed, we argue, in terms of *preparedness*. Although the governance report notes challenges associated with integrating uncertainty into the figure of dual-use, it leaves this thorny question for further investigation.

Containment: Experiments of Concern, Reassurance, and Screening

We have identified three specific types of activities that have been taken up within SynBERC or in direct relationship to the efforts of major SynBERC players. We consider these activities to be exemplary of a shared problem space, shaped by the commissions as described. Specifically, these activities constitute responses to the first of the three problems—safety—identified in the Sloan governance report. Because these activities all serve to anchor the figure of dual-use, they are often taken to be sufficient responses to all three classes of problems, a syllogism we reject. The three types of activities are (1) experiments of concern (EOC), (2) experiments of reassurance (EOR), and (3) procedures for screening synthesis orders.

Understood technically, these three types of activities have become calibrated such that they function as if they shared a common strategic and tactical coherence. Had this coherence and these functions been made explicit, they would correspond closely to what we have called an *equipmental platform*. An equipmental platform produces determinations for how to respond to the figure of dual-use. These determinations are given specific form in relation to the metric of security, the parameter of malice, and so on. In short, this dual-use equipmental platform is characterized by a shared object of interest—experiments and their technical prerequisites— and a shared problem: containment achieved through minimal regulation.

··

Equipmental platforms are characterized by a constantly available generality. Platforms are designed to function effectively in the reconstruction of specific problems, while being plausibly applicable to a range of analogous problems. An equipmental platform can be distinguished from equipmental activities and from specific instances of equipment. An equipmental platform discriminates appropriate (i.e., coherent and cooperable) equipmental activities and functions as the basis for the organization of these activities. The kinds of activities it distinguishes and organizes are those activities that govern objects within a given contemporary figure. These activities taken as an integrated series are instantiated as specific instances of equipment. Put briefly, equipmental platforms function as the basis for the organization of the activities of specific equipment.

Equipmental platforms function in relation to contemporary figures in two important ways. First, platforms contribute to the determination of a problem within a broad field of problematization. Second, platforms contribute to the specification and design of possible solutions to this problem. Equipmental platforms, in short, function as a pragmatic means of transforming aspects (e.g., blockages, difficulties, disruptions of the play of true and false, etc.) of a broader problematization into concrete problems such that these problems can be taken up as a set of possible solutions.

··

Whereas the Fink and NSABB reports and commissions emphasized actors and their intentions, the report on synthetic genomics and governance concentrated more on the type of object to be contained—that is, experiments. Each of the experiments identified in light of the Sloan Report, after all, is the juncture point where the activities of well-intentioned and malicious actors as imagined in the figure of dual-use converge in an object, real or potential. A shift of concern from actors and their intentions to experiments as objects of technical and political intervention functions to allow this equipmental platform to manage the problems of security and preparedness in the name of, and using the strategies designed for, the problem of safety. The intentions of actors, their political milieu, or the ramifications of events, after all, would appear to be amenable to bracketing if we were able to contain the objects that might otherwise be maliciously deployed.

Experiments of Concern (Publications, Capabilities, Community Organization)

Within synthetic biology, experiments of concern—EOCs—were taken up by policy experts and leading scientific players as an obvious place to address the problems raised by commissions and reports. Discussion crystallized around the synthesis of the 1918 flu virus, which had recently been achieved. This technical achievement was framed as an example of both the technical prowess and the potential danger of the increased capacities represented of synthetic biology.

Further, it then became cast as both an ethical issue as well as a technical one. The ethical issue was framed in terms of whether such experiments should be done at all. More importantly, if they were to be conducted—and it was clear that the Department of Defense was going to undertake them— whether that fact should be made public in scientific publications and, if so, in what detail. A distinguished group of editors and scientists, representing the most prestigious scientific journals, achieved a consensus around a set of broad guidelines to guide publishing decisions.

As vague as they were, these guidelines were explicitly intended to be only guidelines; consequently, they were designed not to carry sanctions and were framed in such a way as to avoid legal responsibility for actions taken on the basis of technical materials published in the journals. Anyone intent on ignoring these guidelines would be unconstrained in doing so.

This mode of governance was fully in line with the self-governance principles that the molecular biology community forged at the 1975 Asilomar Conference. The community would apply its own brand of governance. However, the nature of the community, among other significant aspects of the practice of biology, had changed in important ways since Asilomar. To name several of the most important variables: the rise and global spread of the Internet, a much larger community of life science practitioners, much more readily available technology, the rise of the biotechnology industry, perhaps the decline of trust in colleagues, and so on. What might have been plausible in 1975 clearly was not in 2007. This reconfiguration of self-governance attempts to respond to the problem of dual-use by background-ing and bracketing precisely these real-world conditions in relationship to which dual-use has appeared as the principal problem of biosecurity at this historical conjuncture.

A further ramification concerned how much skill, experience, and tech-nology was actually involved in taking sequence information about ex-isting organisms (viruses especially) and turning that information into a functioning pathogen. Some scientists, Roger Brent among them, argued strenuously that the technical challenge was not great and could be met by an ever-growing number of labs worldwide. Others argued to the contrary that, at present, very few labs were actually capable of performing this task, while acknowledging that eventually it would be achievable. The latter ar-gument spoke more to the fact that there was not an imminent emergency than that such capacities were likely to remain out of reach of most biolo-gists, with the exception of a small coterie of highly skilled practitioners. Such highly skilled practitioners would be known to the biological commu-nity and hence would be unlikely to constitute a locus of concern.

This logic led to the proposal that elite technicians and their graduate students be psychologically profiled as to their stability or instability. If only a small group can actually carry out experiments of concern successfully, then the question is not one of publication and circulation. Rather, it is one of the intentions of those few actors who are technically capable and suf-ficiently equipped to carry out malicious actions. The question of technical capacity is thus modified so as to function as another anchor in the figura-tion of dual-use with the parameter of malice.

Given the continued paradigmatic work of the principle of Asilomar community self-regulation, several proposals followed. Drew Endy and others proposed that in order to achieve sufficient and effective self-governance, synthetic biology practitioners would need to formally consti-tute themselves as a professional community, with all that entails in terms of standards of inclusion and exclusion, procedures for governance, as well as codes of conduct. Endy was acting in a proactive manner, since as of 2006 the community of synthetic biologists was small and Endy saw correctly that it would rapidly expand in scale and scope. In addition, such commu-nity formalization would provide mechanisms to establish technical stan-dards for practice. Given the nascent state of this community, the time was fortuitous to formulate pedagogical guidelines in a creative and insistent manner. This technical-pedagogical timing and its attempt to establish an ethos of responsibility and playfulness was especially pertinent given the success of the iGEM competition as a driver for the growth of synthetic biology and its global community. This effort was framed initially in terms of a *culture* of responsible science.

Stephen Maurer was another key figure in this effort to create a formal community of synthetic biology practitioners. Maurer, as we have explained, proposed and argued for two things: (1) a procedure whereby the "community" of synthetic biologists would vote on a set of regulatory controls that would govern the relations of the nascent DNA synthesis industry and the community of synthetic biologists; and thereby (2) a mechanism to monitor "experiments of concern." Maurer emphasized the need to monitor the solicitation of DNA sequences that could be identified by as yet to be developed software as of possible use in known pathogenic agents.

Others, notably Brent, while in favor of developing a culture of responsible science, was vocal in his view that such a soft approach was not by itself sufficient to confront the danger of dual-use. Brent proposed both stronger requirements for individual responsibility and much stronger forms of community policing: anyone who violated the experiments of concern guidelines or who played around its edges would risk having their careers blackballed. And beyond that, there would be criminal sanctions as well, which given the Patriot Act and other legislation was a true threat indeed.

Experiments of Reassurance

While the work and debate about EOCs continues, others began to consider a subsequent set of determinations and a solution to the challenge of an adequate equipmental platform for mitigating the problem of dual-use. We call these *experiments of reassurance*—EORs. Following in the line of reasoning that was successful at Asilomar and subsequent decades, some leading labs such as that of George Church at Harvard, Tom Knight at MIT, and purportedly Craig Venter at the J. Craig Venter Institute have all been attentive to primary safety concerns. They have taken up the challenge of how to genetically engineer organisms that, if they escaped from the lab, would either be harmless or perish. This mode of *containment* and of designed *fail-safe mechanisms* has been an established procedure in molecular biology labs for the last several decades. The experiments that Church is doing could be framed as an experiment of concern, but he understands that and is taking steps to include the exclusions of malicious actions and actors in his designs from the outset. Hence, an attempt is made to transform an experiment of concern into an experiment of reassurance.

Those currently working on the design of synthetic and artificial genomes devote attention and resources to issues of safety and security, and what they take to be attendant social consequences. Their strategy can be called "safety-by-design."[22] There is an explicit effort to design genomes in such a manner so as to maximize control over their functionality. Design attention is devoted to minimizing the risk of survival or reprogrammability outside of the lab. Safety-by-design's purpose is to fabricate genomes that when circulated, have effects, both negative and positive, that can be accounted for and prepared for in advance.

The key *externality* of this approach is that it can only address those aspects of the security challenge that are amenable to technological safeguards. Once again, security issues are framed as problem of dual-use in which the principal challenge arises from the threat of "bad" actors "misusing" technologies created for benevolent purposes. This framing is taken to call for a technological response by existing specialists: Can a biological chassis be designed in such a way that it cannot be subsequently "misused"? Other significant aspects of biosecurity—such as challenges associated with the current political milieu or preparation for unexpected events, which are not amenable to safety-by-design—are backgrounded and bracketed.

To the extent that this externality is taken to be generally sufficient, it becomes a *critical limitation*. That is to say, safety-by-design becomes a critical limitation when it is held that the salient security challenges can be mitigated adequately through technical means, police procedures among and between labs, and trust in the expertise and character of current specialists. Once this externality becomes a critical limitation, there are no other human practices resources within this venue available for responding to other unexpected and unpredicted ramifications.

Screening as Containment: Procedures and Synthesis

At SB 2.0 in 2006, Maurer proposed guidelines for community regulation and screening mechanisms for monitoring "experiments of concern." Simultaneously, there was ferocious behind-the-scenes activity that thwarted the approval of such guidelines. Eventually the Sloan Foundation–funded report on options for governance picked up the main threads of Maurer's proposals, elaborating the modes of minimal regulation of the synthesis industry. The key difference, however, was that this proposal came from within a closed steering group, who certainly attended to a wide range of opinion and advice, but refused to relinquish the framing of the issues and thereby the identification of the community of participants to others.

Over the course of the next three years, several core participants in the Sloan Report contributed to the organization of an international group of gene synthesis companies—the International Association Synthetic Biology (IASB). Unlike the U.S. government, which, in the name of hindering the U.S. biotech industry, seemed reticent to craft regulations based on the Sloan recommendations, a principal rationale for the organization of the IASB was precisely to develop protocols and mechanisms for screening experiments of concern. In November 2009 the IASB proposed a "Code of Conduct for Best Practices in Gene Synthesis," a kind of promissory note that members were committed to inventing and implementing techniques and detailed protocols for screening orders for synthesized DNA, as well as for screening customers. Shortly before the IASB formalized and announced their code of conduct, a second group of synthesis companies—the International Gene Synthesis Consortium (IGSC)—split off from the IASB to form their own consortium. And shortly after the IASB's guidelines were announced, the IGSC circulated news of their own screening protocols. In substance, the IGSC's proposal was much like those of the IASB.

Only after these proposals for self-governance were in place did the U.S. government finally produce a draft proposal for minimal standards for screening. Representatives of both consortia made the appropriate pious remarks in response to the government's draft. Only the NSABB, reiterating the need to protect the freedom and interests of U.S. researchers, seemed unhappy by what amounted to the government's echoing of the industry groups.

The industry screening protocols represent a kind of Asilomar 2.0 for a globalized and security-sensitive world of biological research and researchers. Following the rhetorical and political framings characteristic of the Fink and subsequent reports, the screening protocols take their orientation from the figure of dual-use. Distinguishing themselves from the criminal element, the IASB code of conduct assures us that members of the consortium "strictly follow all regulations and international standards designed to safeguard against intentional or unintentional abuse of synthetic DNA." Synthetic DNA promises revolutionary goods, they hurry to add—"sustainable biofuels, new therapeutics, and biodegradable plastics." Nevertheless, with deep sobriety we must clearly face the fact that today, "the efficiency and potential power of Synthetic Biology can also create the risk of abuse. Through rapid DNA synthesis, bio-risk-associated genes such as toxin genes or virulence factors become accessible to a large number of users."

"It is our declared intention to raise barriers for malign attackers through a number of measures that will combine to protect Synthetic Biology from abuse." It bears pointing out again that an Asilomar-like logic of self-governance and containment remains in play. It bears repeating not because Asilomar is a problem; after all, the safety measures put into place in the 1970s in connection to the Asilomar Conference have basically worked in the way they were designed. It bears repeating, rather, because the logic of self-governance and containment would seem to be an anachronism in a world of global biotechnology. Containment worked for the Asilomar Conference on recombinant DNA because the only serious players were located in government and university labs. As such, physical containment of potentially dangerous biological materials could simply be physically barricaded. Given the fact that "the efficiency and potential power of Synthetic Biology," as the IASB describes it, is already circulating worldwide, how could containment work?

The response from the synthesis companies is twofold. First, synthetic biology, despite its other innovations, still relies on DNA synthesis at a basic level. Second, DNA synthesis is still too difficult and costly to be done by just anyone, and, as such, only a small cadre of trained specialists with specialized equipment can pull it off. In the world of Asilomar, elite U.S. scientists had their government labs with security gates. Synthesis companies have their strategic position as a bottleneck in the global circulation of genetic materials. This bottleneck, we are told, can be turned into a kind of gate at which only the right sort of experimentalist and experiments are allowed to pass through.

> In order to contain the risks of Synthetic Biology and to protect the field against misuse, the Undersigned have adopted this Code of Conduct which provides guidelines for safe, secure, and responsible commercial or non-commercial DNA synthesis.[23]

Containment through the screening of synthesis orders and the street addresses of synthesis costumers is the stated public response. This time containment is a solution that accounts not only for biological safety. It is a solution calibrated to identifying "legitimate" users and uses— which is to say, a solution for the figure of dual-use, with all that entails in terms of security, preparedness, and, of course, malice.[24]

Dual-Use: Ethical Externalities and Critical Limits

As part of synthetic biology, synthetic genomics represents an innovative assemblage of multiple scientific subdisciplines, diverse forms of funding, complex institutional collaborations, serious forward-looking reflection, intensive work with governmental and nongovernmental agencies, focused legal innovation, imaginative use of media, and the like. It is to the credit of the Sloan governance report's authors that in preparing these governance proposals they invited the active participation of diverse individuals. The report's strengths are due in large part to this process.

Although *identifying* a range of security challenges, the Sloan Report *addresses* these challenges in only one frame. At the level of proposed solutions, the report works strictly within a *safety* framework, confronting experiments of concern with screening safeguards. However, we argue that many of the most significant problems related to synthetic genomics cannot be resolved in this way. Whereas a *safety* framework operates within a logic of technological safeguards, a *security* framework additionally attempts to incorporate challenges related to political environment. The options for governance proposed by the report, insofar as they address *security* matters, fold them into screening and licensing technologies. Although acknowledging challenges related to uncertainty, the report offers no concrete proposals for developing frameworks for confronting such challenges. In distinction to *safety* and *security*, such proposals would require a framework of *preparedness*.

Framing the challenges of synthetic genomics as matters of safety, the report recommends development of screening and licensing techniques for controlling who has access to DNA synthesis and for the promotion of "best practices" among scientists. The emphasis is on prevention and protection. The report emphasizes that rogue scientists have ready access to the know-how if not the materials and technologies of DNA synthesis; what's more, these scientists may not form part of the community that would adhere to best practices.

We argue that problems associated with uncertain future events should be framed in terms of *preparedness*. As a technical term, preparedness is a way of thinking about and responding to significant problems that are likely to occur (e.g., a bioterrorist attack or the spread of a deadly virus), but whose probability cannot be feasibly calculated, and whose specific form cannot be determined in advance. In the face of uncertainty, a logic of preparedness, in distinction from containment, highlights the need for vigilant observation, regular forward-thinking, and ongoing adaptation.

Measures for screening and technical safeguards are valuable as far as they go. Given the kinds of problems identified in this report and others, however, it should be clear that these equipmental solutions are not sufficient to address complex contemporary problems. These limitations are reinforced by the fact that dual-use, characterized by the moralism of malice, continues to serve as the figure by way of which problems of biosecurity are taken up.

We are not first-order technicians: we lack funding, organizational infrastructure, and the desire to occupy this slot. After all, the world is full of technical experts vying for funds eager to occupy this first-order position. What we do have to offer are the insights of a second-order positioning: in this case, an analysis of the figure of dual-use. Such an analysis provides some insights into the *externalities* and *critical limitations* of the figure of dual-use as its equipment is being funded, established, and expanded.

Externality 1

The majority of biologists operate as though they were off the hook for the difficult work of responsibility for the ramifications of their work to the extent that their research is (1) carried out with good intentions; and (2) is handed off to other technical experts (e.g., patent lawyers, NGOs, advocacy organizations, venture capitalists, etc.) who will use it to ameliorate health, wealth, or security. Hence, the figuration of dual-use is dependent on the engine of the "good uses" of the biosciences.

Externality 2

Within the synthetic biology community, preparedness is an underexamined category. Vast swaths of industry over the last thirty years have developed risk-management strategists operating in conditions of uncertainty. Synthetic biologists could well profit from acquaintance with the extensive literature and resources devoted to preparedness and risk management in the corporate world.[25]

Just as in the first externality where biologists delegate the biopolitical ramifications to someone else, so, too, in the second. By so doing, the well-intentioned of the world obfuscate the need to transform amorphous and unspecifiable dangers into risks capable of management (however uncertain). Thus the figure of dual-use reinforces a *cooperative* division of labor in which the research community operates more or less in good faith, believing that their work is essentially autonomous. Federal agencies, patent lawyers, safety officers, and ethicists all need to be interfaced with and tolerated, but they can be left to take care of their own business.

As there are always externalities in any practice, pointing them out does not invalidate the practice under consideration. The question is this: What is the price to be paid for the acceptance of these externalities in terms of the capacities that are thereby gained? In our view, the price to be paid for the figure of dual-use is high. The dual-use framing of the problem of security does not, in fact, provide for an adequate response to the range of threats that malicious actors, actions, or unknown nefarious events might bring into the world. In other words, the figure of dual-use involves externalities, which is true of any position or figure. Further, however, the figure introduces critical limitations, which it can't address and devotes energy to obscuring.

Critical Limitation

Once malice and beneficence have been externalized—there are good and bad uses of technology, both of which are beyond the purview of laboratory researchers—a critical limitation ensues.[26] Centrally, the practices that have been established in response to this dual-use figuration actually work against what might seem like an unassailable, if minor, achievement: disciplined self-regulation through minimal training and safety standards. By focusing attention on malicious actors and actions, experiments of concern, and screening procedures, the figure of dual-use actually forces to the side the real question of capacity building.

Practices like codes of conduct, safety procedures, and the like in this figuration become equated with self-improvement. We argue, however, that they are actually a means to cordon off practices that might otherwise transform the practitioners and their community. And hence, since in this figure it is only the Other who needs policing, screening, or improvement, there is no need to devote time and resources for self-improvement in domains other than one's technical methods and know-how.

It follows that the compartmentalization of malice and beneficence is dangerous since real self-improvement "involves a process of de-idealization."[27] De-idealization recognizes that even malicious actors' motivations can be complicated and that well-meaning actors' motivations can also be complicated. The ramifications of either's actions may be discordant with their intentions, even assuming—which we don't assume—that motivations and intentions are transparent. In any case, the attribution of motives is a much more complicated affair than distinguishing good and bad. From a century of depth psychology to major theological debates of the twentieth century through ethical reflections on evil and war, the complex and ongoing debates all point to the fact that a polar typology of good and evil is likely to be misleading and in some instances dangerous.

As long as we don't think more critically and scientifically about the ramifications of instituting a Manichaean division of the world, the figure of dual-use perpetuates a dangerous outside that does not need to be taken into account by the actors involved. If given serious consideration, however, these same dynamics might be reconceived as risks. In that case, preparations might be taken to address some of their ramifications.

The elements that have been drawn together and connected within the figure of dual-use may in fact be relevant variables for thinking through the problem of security and the life sciences today. The task, then, is to work on these variables in such a way that they can become objects of inquiry and practice. Once constituted in that manner, these differentially constituted objects of inquiry and practice can be made visible for questioning in a way that the figure of dual-use precludes. Such questioning might then serve as the basis for reworked equipment—equipment that, at minimum, takes into account the limitations of current practices. This turn would lead to more emphasis on preparedness.

Our work, as we understand it, consists in diagnosis and refiguration of the elements and objects that are being assembled into apparatuses. Such work does not answer the question of how policy makers, governmental regulators, or, for that matter, safety officers should proceed. Rather, our understanding of work is different from these necessary first-order interventions. In our understanding, work is as follows:

> Work: that which is susceptible of introducing a significant difference in the field of knowledge, at the price of a certain difficulty for the author and the reader, and with the eventual recompense of a certain pleasure, that is to say an access to a different figure of truth.[28]

If one is totally satisfied with current conditions, then the call for a different kind of work and a different figure of truth seems unnecessary and a luxury or an obfuscation. If one is not convinced that the figure of dual-use is comprehensive and adequate to the problem of security in the twenty-first century, then a different approach to knowledge and the figuration of security might well prove worth the difficulties involved for those involved in working it through.

9 Toward the Second Wave of Human Practices 2010

Figures of Dual-Use, Biopower, and Reconstruction

> Work: that which is susceptible of introducing a significant difference in the field of knowledge, at the price of a certain difficulty for the author and the reader, and with the eventual recompense of a certain pleasure, that is to say an access to a different figure of truth.
>
> MICHEL FOUCAULT[1]

Our experiment concerns the relations among and between knowledge, thought, and care, as well as the different forms and venues within which these relations might be brought together, assembled, and advanced. Our inquiry is anthropological, a combination of disciplined conceptual work and practice-based inquiry. Our challenge is to produce knowledge in such a way that the work involved enhances our capacities and the capacities of others, ethically, politically, and ontologically. The curve of Western philosophy has been to place more and more emphasis on deciding whether or not something is true while the ethical form of such practices have been neglected or eliminated. We ask: What are the reflective modes and forms for conducting a life engaged with the production of knowledge: the *bios technika*—the arts and techniques of living? In short: What is a worthwhile philosophic and scientific practice today?

Drawing from our experiences in SynBERC, as well as reflections on the state of things in synthetic biology and related domains more generally, we are confident that there is a definite need for a second wave of human practices, concurrent with—but adjacent to—the proposed second wave in synthetic biology. We argue that the initial work required to bring this motion to articulation is figural. To this end, we have been working at identifying and rectifying the elements of a second wave of human practices into objects of thought that are susceptible to being assembled into a figure. The above quote from Foucault elegantly and economically captures this contemporary challenge.

Two reasons inform our decision to undertake the work required for access to a different figure of truth, reasons that we have described at more length in the previous chapters. The first concerns the *scientific stakes* of our undertaking. We have proposed that those involved in human practices, broadly understood, need to shift attention from first-order things to second-order problems and concepts. We noted in chapter 7 that a number of initial proposals have been offered consistent with this shift; these proposals, however, have not yet settled on determinative objects, concepts, or strategies. The labor needed to bring about such determinations—to repeat ourselves—requires work on figuration.

The second reason for the shift from first-wave to second-wave human practices concerns the *ethical stakes* of our experiment. In chapter 6 we proposed that second-wave work requires moving beyond diagnosis of the distinctive characteristics of synthetic biology toward the design and development of effectual and strategic interventions. As of 2010, we remain persuaded that such interventions optimally involve collaboration among and between engineers, biologists, and human scientists, with all such collaboration entails in terms of shared authority and mutual respect and curiosity. Although we have not yet found venues that facilitate such arrangements, we remain committed to helping design and build such venues. To be more precise, our experiences and experiments to date have shown that although SynBERC allows for human practices participants to pose first-wave questions, it does not facilitated second-wave interventions. We remain engaged in SynBERC. Given our inability to affect the habits and dispositions of those we interact with, however, we have decided to devote our efforts to work on the figures of truth shaping synthetic biology and its relations to biosecurity. In brief, work on figuration holds the prospect of enhancing us scientifically and ethically.

The Diagnosis

The challenge is to provide a diagnosis of the contemporary situation of the emergent biosciences and human sciences. We note that although the promise—and justification for funding—of post-genomic biology is largely biopolitical—that is, it contributes to health, wealth, and security of populations and nations—its contributions to the biopolitical domain are at present largely promissory. Of course, discursive contributions can function effectively in a world of symbolic value, and their derivatives can have major ramifications for a surprisingly long time. The biosciences are currently conceived, designed, carried out, financed, and—most importantly—justified with the goal of intervention into natural processes. Some fear those interventions, others applaud them, but partisans of both camps invariably, and for the most part unreflectively, concur that what the biosciences are doing is and should be fundamentally instrumental, guided by a norm of intervention. The opponents of genetically modified organisms, after all, almost never object to environmental cleanup. Currently, this interventionist and instrumentalist norm is embedded in claims to the importance of health, environment, and security.

We have argued that the place in which these practices and discourses have been brought together and given form and backed by powerful institutions is biosecurity specifically framed as dual-use. This situation poses a range of discordancies and indeterminacies for actors and observers alike. The term *indeterminacies* refers to intellectually confused situations that require movement toward increased determination of the problem at hand. We argue that the indeterminacies connected to dual-use turn on the fact that those who give articulation and institutional backing to this figuration are not sufficiently reflective about their own practices and the ramifications of those practices. At the very least, we see this lack of insight on the part of powerful actors as an ethical deficit: the figure of dual-use, which is cast in moral terms, effectively and affectively eliminates what we take to be the ethical complexity of the current situation. Moreover, it reinforces an already existing divide between the life sciences and the human sciences in which the latter is reduced to the role of clearing away those obstacles that might get in the way of an unexamined biopolitical promise as the bioscientific contribution to the human good. While waiting for those contributions to actualize, the bioscientific community effectively assures its own autonomy.

As we will attempt to show, the means to remedy this blockage calls for a strategic reversal. As we saw in previous chapters, within the current situation, spearheaded by mandates proposed by federal commissions, the human sciences have been allotted a definite, if restricted and restrictive, role and responsibility: to advocate and articulate advisory regulatory proposals (largely discursive) nested within the figure of dual-use. We argue that refusing this role is what is needed today: instead what is required is (among other things) a return to the figure of biopower as a step toward more accurately diagnosing the contemporary problem of security, the life sciences, and the human sciences. Given the complex topology of figures operating in the world of security today, we ask: How can they be refigured to provide better scientific and ethical purchase?

Evaluation of the Diagnosis

In our view, the price to be paid for the acceptance of the figure of dual-use is high. The dual-use framing of the problem of security does not, in fact, provide for the possibility of an adequate response to the range of threats that malicious actors, actions, or unknown nefarious events might bring into the world.

As we previously stated, once malice, as well as beneficence, has been externalized—there are good and bad uses of technology as well as good and bad actors, both of which are beyond the purview of laboratory researchers—they build in a critical limitation.[2] Centrally, the practices that have been established in response to this dual-use figuration actually work against what might seem like an unassailable, if minor, achievement: disciplined self-regulation through minimal training and safety standards. By focusing attention on malicious actors and actions, experiments of concern, and screening procedures, the figure of dual-use actually forces to the side the real question of capacity building.

Practices like codes of conduct, safety procedures, and the like in this figuration become equated with self-improvement. We argue, however, that they are actually a means to cordon off practices that might otherwise transform the practitioners and their community. And hence, since in this figure it is only the Other who needs policing, screening, or improvement, there is no necessity to devote time and resources for self-improvement in domains other than one's technical methods and know-how.

It follows that the compartmentalization of malice and beneficence is dangerous since real self-improvement "involves a process of de-idealization."[3] De-idealization recognizes that even malicious actors' motivations can be complicated and that well-meaning actors' motivations can also be complicated. The ramifications of either's actions may be discordant with their intentions even assuming, which we don't assume, that motivations and intentions are transparent. In any case, the attribution of motives is a much more complicated affair than distinguishing good and bad. From a century of depth psychology to major theological debates of the twentieth century through ethical reflections on evil and war, the complex and ongoing debates all point to the fact that a polar typology of good and evil is likely to be misleading and consequently dangerous.

The biologists and engineers operate as though they were excused from the difficult work of responsibility for the ramifications of their work to the extent that their research is (1) carried out with explicit good intentions; and (2) is handed off to other technical experts (e.g., patent lawyers, NGOs, advocacy organizations, venture capitalists, etc.) who will use it to ameliorate health, wealth, or security. Hence, the figuration of dual-use is dependent on the engine of the "good uses" of the biosciences. By so doing, the (self-nominated and self-evaluated) well-intentioned of the world thereby obfuscate the need to transform amorphous and unspecifiable dangers into risks capable of management (however tentatively), if one is to act with foresight and maturity. Thus the figure of dual-use reinforces a *cooperative* division of labor in which the research community operates in what they take to be good faith and believes that their work is essentially autonomous (e.g., the federal agencies, patent lawyers, and safety officers all need to be interfaced with and tolerated, but they can be left to take care of their own business).[4]

As long as we don't think more critically and scientifically about the ramifications of instituting a Manichaean division of the world, the figure of dual-use will continue to perpetuate a dangerous outside that does not need to be taken into account by the actors involved. If given serious consideration, however, these same dynamics might be reconceived as risks. Some of the elements that have been drawn together and connected within the figure of dual-use may in fact be relevant variables for thinking through the problem of security and the life sciences today. The task, then, is to work on these variables in such a way that they can become objects of inquiry and practice. Once constituted in that manner, there is the possibility of questioning their pertinence in a way that the figure of dual-use precludes.

Security Figure 2: Biopower

Given this situation, what is to be done? Strategically, a first move consists in assessing, diagrammatically, the relations among life, security, and *anthropos*—elements currently being configured as dual-use. One way of beginning a critical analysis is to return to the figure of biopower as brought to articulation by Foucault. One reason, only partially arbitrary, for returning to Foucault's discussion is that he named his elaboration of biopower "security." Of course, we are not advocating a return to an eighteenth-century view of these problems and their solutions, only that seeing a different formation of elements may enable us to gain clarity on our current blockages.

Foucault's lectures on "Security, Population, Territory" provide trenchant, if only partially adequate, tools for orientation within our contemporary situation. For current purposes, therefore, it is sufficient to present here a minimalist schematization of Foucault's concept of "security." Using Foucault's analytic in this manner enables one to "enter into" (to use a term of Niklas Luhmann's) the amorphous contemporary, and begin to explore and organize it.

At the very least, a return to a formulation of the elements of the figure of biopower will underscore that whatever else the figure of dual-use may be, it is not a figuration of biopower. This contrast opens up a space of comparison: malice, for example, plays no role whatsoever in the emergence of the figure of biopower. Quite the contrary, the technicians and functionaries struggling to cope with recurrent famine, urban revolt, and epidemic outbreaks sought above all to eliminate or downplay existing moral framings of these phenomena so as to arrive at a more discerning and operational set of understandings and tools for effective intervention. Our diagnosis, as well as our strategy today, begins in a similar mode.

Object: Milieu

Although Foucault's 1977–78 series of lectures at the Collège de France — "Sécurité, Territoire, Population"—is best known for its introduction and initial elaboration of the concept of "governmentality," the first three lectures contain a clarifying discussion of the terms "population" and "security."[5] Consistent with his genealogical method tailored for a history of the present, Foucault's challenge is to specify sites of emergence. Hence, as is his habitual want, he provides the genealogical elements for a history of the present but not for the further elaboration and inquiry that would be required for an anthropology of the contemporary.[6]

Foucault first addresses the question of which objects each apparatus is designed to take up and to work over. *Sovereign* operates on a *territory*. *Discipline* operates on *individuals*. *Security* operates on a *population*.[7] That being said, each of the technologies must also take into account, albeit in a dependent manner, the other two variables (hence the title of the course). Thus, for example, while the main target of disciplinary technology is individuals, it operates within a space (part of a territory) that it seeks to isolate out and to homogenize. Within such a space, individuals are distributed, fixed, and observed, but only after they have been extracted from an existing multitude (population). By definition, a space of security, as opposed to one of discipline, is neither an empty space nor is it worked over in order to produce uniformity. Rather, a space of security is one that contains heterogeneous material givens (both natural and social) that must be taken up operationally in accordance with their own inherent qualities (thus preserving specificity of difference). A security apparatus does not seek to totally refashion these givens; rather it operates with a rationality of maximization of positive elements and minimization of negative elements. To arrive at a point where such maximization and minimization can be comprehended, knowledge of the material givens must be constantly accumulated and put under evaluation. Thus, in a security apparatus, its elements, these material givens, are the target of *modulation*, not total reformulation.[8] Security is always in motion. Its traditional goal, after all, is the best possible circulation of goods, things, and people.

While discipline works on the *present* to shape a controlled, stabilized future, security does not aim at *future* developments that are fully controllable or regular. One way that it does this is to be attentive to the history of individual elements, apparatuses, and assemblages. Hence, deploying the logic of total control of a *tabula rasa*, as in discipline, would be inappropriate and counterproductive. Finally, a security apparatus takes up the problem of how to manage an indefinite series of elements that are in motion. This motion is understood within a logic of *probable events*. As Foucault, Ian Hacking, and others have shown, statistics and probability were born as parts of an art of government in the seventeenth century. The new mathematics was a corollary of a new mode of attention to things. To understand an event set within a probability series, it was mandatory to develop means to be attentive to the previous configurations (of these elements and series and their recombinations), their present state, and their potential reconfiguration.

The general name given to the space of security, the site of this probability-inflected series of events, is the *milieu*. A milieu is both the environment within which actions take place and the material of which action operates. *The milieu is the dynamic ensemble of historically interwoven natural givens and artificial givens.* A security apparatus must identify and intervene "precisely there at the point where there is interference between a series of events produced by individuals, groups and populations and the series of events of a quasi-natural type that are produced around them." Within the milieu, there is a cybernetics of cause and effect. Said another way, the milieu is the sum of a certain number of "*effets de masse*" weighing on those who inhabit it and seek to shape it.[9] By definition, a milieu is dynamic and reflexively transformative. Foucault concludes this lecture with this extravagant claim: It is within the rationality of the milieu as a target of a security apparatus that the "naturalness" of the human species makes its appearance.

In the second lecture (January 18, 1978), Foucault takes up the security apparatus from the point of view of *events*. Security is a centrifugal device. It constantly seeks to integrate, in its own manner, new elements. Milieus tend to be expansive and are not stable. Whereas discipline seeks to homogenize and order everything within its orbit, security lets diverse things go as they have been going as long as it remains possible to observe them and, if need be, modulate them. A security apparatus seeks to apply itself to existing details of existing processes so as to be able to intervene in the course of future events. Whenever possible, security does not seek to block the course of things or to forbid actions. It is positioned at a meta-level, so as to ascertain first where things are going regardless of whether one approves of that directionality. Security and this mode constitute a major innovation, resisting the application of a priori judgments, especially moral ones. Security operates in an optimal manner when it succeeds in taking things up at the level of their "effective reality." Its challenge is to use existing elements in a milieu to annul or limit nefarious tendencies and to maximize beneficial ones through means appropriate to their specific conditions and potentials.[10]

We can see that the figure of biopower with its attention to milieu, population, modulation, anti-moralism, temporal specification, events, and probability provides an arresting contrast to the figure of dual-use. We should add in passing that the figure of biopower equally stands in stark contrast to other figures at play today such as humanitarianism, which is actually closer to dual-use than to biopower.

Security: The Limits of Biopower for Synthetic Biology

A constant refrain from advocates of Mode 1, as well as from the funders of synthetic biology, is that the principal focus of human practices should be biosecurity. The hoped-for outcome of this work would consist of guidelines for the regulation of synthetic biology in such a way as to minimize adverse security events in the contemporary political milieu. The question is this: To what extent is such an approach appropriate and worthwhile?

To date, Mode 1 work on biosecurity has been framed in terms of the challenge of risk assessment. What kinds and levels of risks does synthetic biology present, and what kinds of regulations can be put in place in view of those risks? A central—though insufficiently examined—supposition of Mode 1 experts is that they are capable of conceptualizing dangers, both known and unknown, as risks that can subsequently be assessed with appropriate rigor and plausibility. In our view, the existence of such a capability is very much in question.

As we have previously suggested, following Niklas Luhmann, it is useful to distinguish dangers from risks.[11] Dangers are empirical factors that exist in the world in a scientifically under-examined state. In Luhmann's formulation, dangers are inchoate, which is to say they lack a conceptual structure to order them, and hence they do not operate in the domain of legitimate truth claims. It is only once dangers are conceptualized and enter into a grid of knowledge that technically they become risks. Risks, unlike dangers, can be scientifically assessed. The relative likelihood of a series unfolding in a particular manner can then be determined, and strategies for minimization and maximization elaborated.

It is clear that synthetic biology, to the extent that it makes biology easier to engineer, introduces new dangerous actions and actors into the post-9/11 political milieu. The question of whether or not, as of today, such dangers can be assessed as risks, however, remains an open one. That question depends in part on the ability of Mode 1 specialists to figure synthetic biology through the terms of biopower, that is, as a field of probabilities made available to techniques of normalization.

There are, of course, different ways of thinking about dangers, and therefore different ways of turning them into risks. The biopolitical mode is probabilistic. On the most refined readings, security-related events in synthetic biology are conceived in terms of the ratio between probability and consequence. Events of interest are taken to be of low probability but of high consequence. The questions are this: How is such a judgment made? How can it be known whether or not the probability of events is low or high? Such judgments, again strictly speaking, can only be made in relation to a multiplicity of actual events. It is only when there have been multiple events categorized and classified that they can be taken up as a series. It is in relation to this series that claims of probability can be systematically derived.[12] Specific sectors of these probabilistic series could then be assigned risk values and consequences could be estimated.

The challenge for Mode 1 experts is not only to establish probabilistic series but also to formulate techniques and technologies—equipment—for intervening in those series. Here normalization enters in. Probabilistic series can only be established within a structured field of relations. This means that a metric is needed for determining what things qualify. Once that metric is established, then things can be picked out as elements, associated, displayed, and coordinated as a relational field. In order to achieve the end of security through regulation, the elements picked out will need to have the minimal characteristic of being susceptible to being normed.[13]

Mode 1 experts in synthetic biology are not yet capable of converting dangers into risks. Indeed, they do not yet have standards for picking out what counts as dangerous with any analytic power. Many of the critical limitations of prior figurations can certainly be assessed, as we have shown in our analysis of research programs and human practices modes. It is plausible to hold that probabilistic series that would permit the determination of risks in synthetic biology may never be produced. If commonly accepted speculations about dangers prove correct and security events involving one aspect or another of synthetic biology are of the type "low likelihood/high consequence," there may never be a minimum number of actual events required in order to create a series in relation to which high and low probabilities can be calculated.

In the meanwhile, there are important initiatives under way outside of

SynBERC that attempt to compensate for this lack of actual events by creating hybrid modes that straddle Mode 1 and Mode 2. For example, members of the EU research initiative Synbiosafe are conducting hundreds of interviews with biologists concerning their views of safety and security.[14] One of the aims of these interviews is to determine the relative conscientiousness or awareness of biologists relative to questions of security and synthetic biology. The data they are amassing is certainly susceptible to verification and, as such, can be made into a field of normalization. Such a field, however, will not be useful for the regulation and minimization of security-related events, per se. Rather, it will be useful for establishing norms and distribution values relevant for designing educational and other consciousness-raising equipment. In other words, the verified reductions produced by the Synbiosafe researchers will contribute to the problem of security only indirectly: they will tell us the extent to which current biologists take the problem of safety and security seriously. Presumably, increasing the number of biologists who do take these issues seriously will affect the likelihood of a dangerous event occurring, although such likelihood— to make the point again—still could not be calculated.

Much of what is happening both in synthetic biology (more generally), as well as in SynBERC, exceeds the figure of biopower. What this means is that SynBERC is a venue at odds with itself twice over. In the first place, it draws its official self-understanding and organizational model from the predominant synthetic biology manifesto—the parts-based approach. Meanwhile, the actual research programs in the SynBERC laboratories (and other synthetic biology laboratories) are ramifying in ways that diverge from the ontological suppositions of that manifesto. A similar and related tension marks human practices. If problems ensue when one attempts to use previous sets of ethical equipment on objects for which they are not suited, then the unresolved ontological question contributes in a significant fashion to a state of ethical stasis. Ethical *stasis* under the present conditions might well be relatively harmless—mere incoherence—or such stasis might contribute to a situation in which dispositions, forms, and practices are maximized and strengthened that will obscure the identification of problems and consequently contribute to blocking or foreclosing the pathways to remediating contemporary figures.[15]

Reconstruction: Toward a Figure of Ethical Practice

The challenge is how do we remain close to the biopolitical goods of health, wealth, and security without simply aligning these goods into a grid of normalization applied to populations to be governed? That is to say, what ethical practices are available today given a reconfiguration of these biopolitical variables? We are aided in this effort by the further contrast of the figure of dual-use, which involves an explicit, if dubious, moralization of the problem. We ask again, given the complex topography of figures operating in the world of security today, how can they be refigured to provide better scientific and ethical purchase?

In the remainder of this chapter, we are pursuing the hypothesis that by taking the figures of dual-use and biopower as part of a (virtual) series, we can at least begin to specify the variables for a reconstructive ethics of the contemporary. Our argument and subsequent inquiry turn on the twin assumptions that (1) neither dual-use nor biopower in the mode of intervention of normalization of populations can provide an ethics; and (2) that an ethics is a beneficial orienting, diagnosing, and guiding ethos for articulating a form of life in which flourishing rather than amelioration or maximization is privileged.

A prominent placeholder candidate to occupy this philosophical slot is the term *reconstruction*. We take the term *reconstruction* from John Dewey, specifically his use of it in the foreword to his book *Reconstruction in Philosophy*. Since at least 1920, Dewey had deployed the term to guide his inquiry into the good of thinking understood as a practice in the historical world: What is *reconstruction*?

> Reconstruction can be nothing less than the work of developing, of forming, of producing (in the literal sense of the word) the intellectual instrumentalities which will progressively direct inquiry into the deeply and inclusively human—that is to say moral—facts of the present scene and situation.[16]

That challenge, the challenge of reconstruction, while having a certain generality as Dewey formulates it, has, in his eyes, an urgency under conditions in which the technical accomplishments of science were expanding as well as separating from the older moral base in which it was held they used to be embedded. However contestable the last assertion may be, Dewey's diagnosis retains an actuality that would be hard to gainsay. Dewey's assertion that modern science previously was embedded in a satisfactory moral form of life is open to question. That thinking had been given different forms and figurations in which truth-seeking, truth-speaking, and subject formations were recursively aligned, however, is not a controversial claim.

Second-Wave Figuration: Dual-Use, Biopower, Reconstruction

Figure	Site of Intervention	Mode of Intervention	Mode of Veridiction	Object
Dual-Use	Techno-structures	Containment	Malice	Subjects
Biopower	Milieu	Normalization	Probabilistic (series)	Population
Reconstruction	Milieu	Techno-cosmopolitan	Warranted Assertibility	Form of Life

Ethical Series: Milieu, Intervention, Veridiction, Form of Life

The term *reconstruction* is directed at the need and desire to develop and practice an ethical form of inquiry. Its object—that which it is concerned to identify and remediate—is a *form of life*. Such an object is not ontologically given but attains its existence and takes its form through a recursive process of inquiry, judgment, and further inquiry. It can, therefore, also be described (following Dewey) as an *objective*.

In the two other figures under consideration here, a similar logic applies: for dual-use, the object and objective are to identify and separate good and bad *subjects* and their practices. In the figure of biopower, in order to efficiently and effectively normalize a *population*, it must be identified, analyzed, and observed with appropriate modulated interventions, again as both the object and objective. In both figures what we are calling the ethical form of reconstruction is precluded. In dual-use, moralism is fixed from the start and the challenge is to identify empirical examples fitting one or another of the binaries, good or bad. This form precludes self-improvement, critical reflection, and ultimately ongoing inquiry. In the figure of biopower, the object is a population to be normalized. As we have just discussed, this distributional operation was originally designed against the prevailing moral judgments, which, its inventors held, blinded or obscured the reality that they sought to manage. As our goal is a form of reconstructive practice, the critical limitations of the figures of dual-use and biopower concerning the form of inquiry into their selected objects need to be remediated.

The *sites of intervention*—techno-structures, historical-natural milieus, milieus in the process of refiguration and reassemblage—vary as well. In dual-use, the figure is constructed so as to constrain and contain relevance to a *technical* level. For biopower, intervention on the population set within its *milieu* should be as analytic and rigorous as possible given the nature of the laws held to be governing vectors such as climate, market forces, plant physiology, disease, and the like; older moral ideas, it was held, tended to

obscure the natural functioning of these vectors. That being said, over time
a certain moralism of the naturalness of things, an ethic of laissez-faire,
itself became a moral force that blinded certain practitioners to the ramifications of their interventions in the *milieu* that concerned them. In the
figure we propose to develop in this chapter, the site of intervention is still a
natural-artificial milieu; the challenge, however, is to introduce reconstruction through active inquiry and ethical intervention.

The *modes of intervention*—containment, normalization, techno-
cosmopolitan—also vary. As we have seen, the mode of intervention of
dual-use is *containment* and that of biopower is *normalization*. As concerns
the mode of intervention of reconstruction, we find here an interesting
overlap with the site of intervention of biopower as well as the second
wave of synthetic biology. One of us (Rabinow) coined the term "techno-
cosmopolitanism" to designate a mode of planning and to differentiate it
from the high or middling modernism in which planning proceeded on
what it took to be a blank slate.[17] The mode of intervention of biopower as a
form of techno-cosmopolitanism proceeded on the basis of need-to-know
specifics and to find technologies of modulation of those specifics that were
economical.

Techno-cosmopolitanism can be defined as the attempt to regulate history, society, and culture by working over existent institutions and spaces—
cultural, social, and aesthetic—which had been seen to embody a healthy
sedimentation of historical practices. Its technological operations were applied to specific customs, cultures, and countries—hence, cosmopolitan.
For example, technical dimensions of urban planning in Morocco would
resemble those in Brazil (both would have a small number of zones, a circulation system, sewage treatment, and so on), but the well-planned city
would artfully integrate and strengthen topographic, cultural, and social
specificities into its plan. The art of urban planning, and thus of a healthy
modern society, was held to lie precisely in this orchestration of the general
and the particular. In this scheme, modernization must be understood as
the identification, evaluation, and regulation of tradition in the name of
productivity, efficiency, and welfare.

Remembering that a term consists of a word, a concept, and a referent,
we can see that in the second wave of human practices, the referent (at the
very least) of techno-cosmopolitanism differs from both its previous uses
as a mode of urban planning and from its deployment as a design strategy
in the second wave of synthetic biology. In the second wave of human practices, we are still working here at the figural level.

A first challenge of establishing this mode of intervention has been to
identify and seek to construct a venue in which such work can proceed. Our

first-wave experiences in SynBERC convinced us that if there was to be a reconstructive venue of human practices in synthetic biology, it would not be located within one organization, that the range of issues involved would be distributed broadly and differentially, that the ramifications of insights, discoveries, and blockages would have to be monitored and pursued in a dynamic fashion. In this light, the very shortcomings of a particular organization or its actors can point to a questioning as to whether these are more or less accidental deficiencies, initial stages in a recursive process, or catalysts for others, elsewhere, to do better.

The *mode of veridiction*—malice, probabilistic series, warranted assertibility—likewise vary. In the figure of dual-use, one knows one has arrived at a satisfactory truth claim when one can identify and *classify* a good or a bad subject or practice. In the figure of biopower, only those claims count as true or false that can be arrayed according to a *probabilistic series*.

The speech acts that can be authorized as true and false in reconstruction are those assertions that can be put to the test in experimental and pragmatic situations and subsequently can be reused in a reworked form. These experimental and pragmatic situations are more than just laboratory exercises: they are discordant and indeterminate situations that have occasioned thinking and inquiry. As such, although technical virtuosity and prowess may contribute to the growth of capacities in thinking, such capacities only count as warranted assertions when they are both the product of a stage of inquiry and can be counted on as reliable in the design of the next stage of the inquiry.

Design: Parameters and Determinations

The process of inquiry into contemporary figures involves and requires staying in the midst of things of the world but of transforming them in specific ways so as to give them the kind of form as objects and objectives that is determinate. Therefore, our work in 2010, having passed through four years of experience and diagnosis of the first wave of synthetic biology and human practices, consists in moving through that diagnosis so as to identify and formulate the parameters for a second-wave experiment in human practices.

In light of our initial experimentation, we decided to take up our challenge less as observing actual figures coalescing in the world, and more as the identification of the elements, objects, and parameters of a virtual figure in need of form. A virtual figure functions as a series of scientific and ethical design parameters and modes of composition. The design parameters facilitate critique and construction by throwing into relief *externalities* and *critical limitations*, and by showing where and how those externalities and critical limitations are contributing to scientific indeterminacy and ethical discordancy. They facilitate clarification, more precise adjacency, and a sharper orientation to secession in relation to existing configurations of indeterminacy and discordancy.

Such parameters serve to delimit and inform preliminary designs for a practice of further inquiry. The designation of these parameters leads to the specification of determinations. Such determinations are a significant step more concrete than the parameters that oriented us to the form of the figural challenge. Hence the purpose of design does not lie in its ability either to produce tools that represent a preexisting situation or in its ability to construct an entirely new one. Rather, design of parameters for inquiry facilitates reiterated and controlled adjustment of existing elements as well as emergent ones so as to arrive at a determinate and concordant situation— at least figurally.

Parameter 1: Figuration

One of the genre characteristics of first-wave synthetic biology has been framing. Such framing contributes to the production of a *figuration* that facilitates orientation and the construction of overarching narratives of the field's constitution and promised future horizon. Dual-use and biopower have been the predominant figurations during the first wave of synthetic biology. In the second wave, having argued above that neither of these is capable of providing an adequate diagnosis or ethics in the sense of a practice of care and knowledge, we decided that the task is to assemble these figurations into a series, which we have done. The reason for this is that we are not dealing with an epochal change, but a virtual refiguration.

Parameter 2: Milieus

A challenge for second-wave synthetic biology consists in moving beyond existing figurations insofar as they remain "too simplistic to efficiently create complex biological systems and have so far limited our ability to exploit the full potential of this field."[18] One proposal in the second wave of synthetic biology has been to shift from parts and minimal modules as the primary units of concern to domains of biological complexity that hold the promise of being manageable. Biologists call these domains *milieus*. The figure of dual-use produces systematic blindness about the self and others, as we've outlined in the previous chapter, insofar as their motivations and capacities cannot be reduced to binary moralistic terms. The figure of biopower lacks both hermeneutic and ethical holding points to identify signification and critical limitations. In the second wave of human practices, a challenge is thus to take account of these two existing figurations as elements in a broader and more complicated milieu, whose figural articulation constitutes a key objective of the second wave of human practices. This work consists in attention to specific zones in which interfaces among and between elements of the previous and emergent figurations could be given form.

Parameter 3: Capacity Building

Capacity building designates a mode of strategic intervention calibrated to identifying and regulating a limited set of existing functions within a milieu. In the first wave of human practices, the capacities to be identified and specified are diagnostic ones. We were convinced that unless we had an analytically powerful diagnosis of the current situation, we would be proceeding blindly. In the second wave of human practices, it follows from our discussion of milieu that the challenge, both figurative and eventually practical, is to specify and enhance those capacities capable of functioning in a reconstructive manner.

Parameter 4: Leveraging

Leveraging is a mode of composition that takes advantage of existing talent, technology, and resources, adjusting their interfaces such that the resulting connections should increase capacities in a way that could not have been the case had these elements been taken up serially. In the second wave of human practices, having identified specific capacities, the challenge is one of remediation. Remediation consists in identifying weaknesses and insufficiencies within those zones within the current milieu that are open to strategic intervention, at least figurally.

Turning such a series and figure into a practice is an even more challenging task. The challenge is yet more daunting if one does not have the sanction and resources of the state (whether that of eighteenth-century France or twenty-first-century America) authorizing, legitimating, and supporting one's labor. For current purposes, the identification of a problem—why does figuration seem actual?—is already an accomplishment; an accomplishment that brings with it a certain pleasure and the promise of difficulties to come.

10 Lessons Learned 2010

From Indeterminacy to Discordancy

In March 2010, Rabinow was removed as the head of Human Practices.[1] It was announced that a social scientist with expertise in "bio-risk" was to be hired, and that the name of Thrust 4 was to be changed. This event, and the episodes leading up to it and immediately following it, definitively transformed our relationship with SynBERC from its inception in early 2006. We present here a brief chronicle of these crucial events: effectively the end of a second phase of our experiment. We include passages from official communications that constitute the documentation of these episodes so as to make them available for the record, and we provide synoptic conclusions about the organizational hindrances to the success of the overall experiment.

SECTION I. ADMINISTRATIVE POWER PLAY:
LESSONS LEARNED

Out of the blue, in March 2010 Jay Keasling informed us by e-mail that the NSF had decided to remove Rabinow as head of Thrust 4 and that the name *human practices* must be replaced by a name that highlights work on "bio-risk." This decision came from the office of Lynn Preston, leader of the ERC Program and Deputy Director of the NSF's of Engineering Education and Centers Division, as part of the official report from the fifth annual site visit review. We do not know how this decision was reached, as alternate accounts of the events have circulated. We have only the account provided in the official report, which we quote here verbatim:

> *The presence of a Human Practices (HP) Thrust represents a significant opportunity to develop a coherent policy for the use of synthetic biology in a safe and effective manner that considers appropriate biological and public risks and ethical issues. The human practices approach also represents an opportunity to address some of the major IP issues associated with open sourcing of select materials*

along with appropriate protection of key inventions. Unfortunately, the Human Practices Thrust appears fragmented and not fully productive in achieving the major objective of developing an effective public policy for synthetic biology. Instead, some components of the HP research appear to be primarily observational in nature rather than proactive and developmental. A major redefinition in some of the current work under way in this thrust along with a change in leadership will be necessary to facilitate significant progress prior to the sixth-year renewal evaluation. The IAB [Industrial Advisory Board] also expressed concerns about the lack of a human practices program that is useful to industrial participants in synthetic biology. It is hoped that with new leadership and a more focused agenda, the human factors t[h]rust can contribute policy and awareness of risk, biosafety, biosecurity, and ethics issues that will be infused throughout all Center activities, from outreach, to education, to research in each PI's laboratory.

Continued support and the requested increase in funding for year four is recommended, contingent upon the submission to NSF of an effective strategy to address deficiencies in leadership of Thrust 4 and its integration with the rest of the ERC. The plan for the leadership change is due within 30 days from receipt of the site visit report. It will include a plan for searching for a new Thrust leader, the timeline for that search, and the appointment of an interim thrust leader. The Director will develop this strategy in consultation with the other thrust leaders and faculty, the industrial and scientific advisory boards, and others as appropriate. It is recommended that consideration be given to finding a person with experience in biotechnology policy, the biological risks, safety and security issues in synthetic biology and/or genetic engineering, and the ability to communicate these issues effectively to research communities outside social scientists, policy makers, and industry. The SVT further recommends that the name of Thrust 4 be changed to highlight an increased emphasis on biological risk in synthetic biology. Upon receipt of this plan, the NSF Program Director will determine the quality of the plan and its responsiveness to the issues raised before recommending continued support.

This recommendation is based on the following: (1) the Thrust 4 research agenda is fragmented and does not prioritize the range of issues around biological risk, biosafety, and biosecurity and IP as strongly as it should; (2) the array of expertise in Thrust 4 as currently organized needs to be augmented for the new priority; (3) much of the extant work in Thrust 4 is of high quality and needs to continue in order to provide the capacity for ongoing and improved collaborations with the new priority and across the Center; and (4) the skills base of the current leader of the Thrust, Professor Rabinow, is not aligned with the expertise required by the new agenda. While the site visit team appreciates the contributions of the leader of Thrust 4, Dr. Rabinow, to the early development of

SynBERC, a new skill base for the leader is needed to fully engage the SynBERC team in assessing due to the formative stage of this technology. The biological risks associated with synthetic biology, integrating this knowledge into their research, and developing synthetic biology design guidelines. This recommendation is consistent with NSF's original guidelines for addition of this thrust in SynBERC and is a natural progression of the development of SynBERC, the field of synthetic biology, and the expressed need of the IAB for clear and actionable guidance.[2]

Context 2009–2010

This set of recommendations was a surprise in its timing, but not a surprise in its strategic implications. The 2009 site review assessment of the work and accomplishments of SynBERC Human Practices had been extremely favorable. That official report summarized the purpose of Thrust 4 as the challenge of inventing strategies for cultivating "a new ethos of responsible scientific and engineering practices," and for assessing the ways in which synthetic biology is "both an opportunity for even greater flourishing and a threat to existing modes of well being." In this light, the 2009 report recognized both that our work to date—characterized as "vital, informed, high-level research initiatives"—could provide a model for work in other similar centers. The overarching challenge, as articulated in the 2009 report, was to "create a new kind of research culture where scientists and engineers are working with scholars in the humanities and social sciences to proactively identify and address ethical and social issues integral to their own research."[3]

Collaboration of a serious and inventive kind, in short, remained the central focus and mandate for Human Practices. Moreover, the principal blockage to such collaboration, according to the site visit team (SVT), fell on the biologists' and engineers' side of the ledger: "Science/engineering researchers do not seem to fully appreciate that there is an opportunity—of the same kind as in other technical areas—to provide world-class leadership in an influential, emergent area of [human practices] research."

The 2009 site visit team's assessment cohered with our experience, and with the findings of human science researchers in other synthetic biology venues. Many biologists and engineers remain resistant to active collaborative work on ethics. "For many researchers," as the SVT asserts, "there still seems to be a kind of defensiveness and too close of an association between ethics and externally imposed rules or constraints." Such a diagnosis is confirmed by studies conducted by researchers at the European Synbiosafe project, which report that many bioscientists and engineers admit to not actively considering questions of ethics as part of their research.[4] Follow-up studies go further, showing that many researchers claim that there are no major ethical concerns in synthetic biology, and that ethical concerns are only a matter of "public perception."[5]

Moreover, it bears underscoring that during the formal sessions at the third-year site review, the social science members of the site visit team vigorously defended the need for Human Practices to engage in second-order observation of the unfolding bioengineering program and to conduct anthropological inquiry into the question of what responsible practice in biology means today. After all, they stressed, the premise of SynBERC is to make biology easier to engineer and thereby available to a much wider range of practitioners. To the extent that these goals are realized, prior modes and venues for regulating biology—on ethical, safety, or security fronts—could no longer be counted as sufficient.

In light of these exchanges, and with the 2009 report in hand, we cast our research protocols for 2009–10 in strongly anthropological and inquiry-based terms, taking ethics, DIYbio, security, and preparedness as core orienting themes. Further, by September 2009 we had fashioned a coherent framing for such an inquiry-based approach, a framing summarized in the second half of this volume as the "second wave of human practices." Although such a second-wave approach was informed by our insistence on adjacency and secession as a productive position relative to the first-order stakes of the biological and engineering work, it was also resolutely informed by our commitment to find a way of shifting from second-order observation to second-order strategics. Whatever else a commitment to Mode 3 research entails, it does not include abandoning direct participation in the core mission or stakes of synthetic biology or SynBERC.

In the light of all this, and in good faith, we prepared for the 2010 site visit with a certain optimism informed by a sense that our second-wave framing of human practices was generative. Given the ten minutes allotted to our presentation of the prior year's work, we settled on a presentation that provided a concrete demonstration of the diagnostic emphasis of our first three years' work as well as the strategic and inquiry-oriented character of work now under way. This approach and presentation was approved by SynBERC authorities, as they were keeping a close watch not only on the structure and format of the site review, but of the timing and content of the presentations as well.

The presentation at the site visit appeared to go well. Informal feedback from SynBERC officials was extremely positive. Likewise, the SWOT report questions contained only one pertaining to Thrust 4. Moreover the question appeared to emphasize continued work with more focus, but nothing larger than that. The Thrust 4 PIs responded with a certain measure of relief.

Observations: Insufficiencies

We had been told multiple times (by engineers and biologists in SynBERC) that the head of the ERC's directorate, like most engineers, would be disinclined to understand anything too complicated. Those who offered these observations did so as friendly advice to us not as a criticism of engineers, underlining that it would be wise for us to lower our expectations and adjust our responses. However, as other engineers at the NSF displayed curiosity and openness while having the fortitude to admit that they had a lot to learn from other disciplines, we were encouraged to continue as we saw fit in the first three years of SynBERC. Hence, we were not dealing with foreclosed possibilities of disciplinary and organizational structures, only dispositional tendencies and the willful actions of specific actors.

That bio-risk and biosecurity were emphasized by the NSF ERC Directorate is hardly surprising; these and related issues, after all, had been high on the agenda of concern in general in government circles. That these were cited as the designated topics of our insufficiencies, however, was surprising and at odds with the facts of the situation. One of us (Rabinow) with colleagues had interviewed and consulted with people (some in high official positions) in Washington, DC, for several years during the course of their project on "Global Biopolitics of Security." We knew that there were knowledgeable observers and officials in Washington who articulated a complex understanding of the security situation, including: (1) the need for a preparedness strategy (in addition to safety safeguards) and, most relevantly here, (2) the necessity of having sophisticated social science input into any understanding of such issues. It was in part starting with that experience as background that Thrust 4 proceeded as it did, and projects on security and preparedness had formed a core part of our research from the outset. Our publications and the publications of our colleagues on these topics were made available to the NSF ERC Directorate and the SynBERC leadership.[6] We conclude that the refusal on the part of the leadership of the NSF engineering directorate in 2009–10 to recognize or credit our framing of these issues, and the concurrence of our work with other efforts in Washington, is at best ignorance, at worst negligence.

We had given presentations at all the previous site reviews, included materials in the written reports, and provided posters explaining dimensions of these understandings of the security situation. We connected them to the deliverables that we were being asked to provide and attempted to articulate why we (following others in the field) were taking the approach we did rather than answer what we considered to be the wrong questions. As far as we can tell, there was a refusal on the part of Lynn Preston and key members of the Industrial Advisory Board to ever seriously consider our reasons for working on one set of problems rather than another. Significant parts of our approach had appeared in peer-reviewed journals including *Nature Biotechnology* and *Systems and Synthetic Biology* and were being acknowledged as innovative by others working alongside us in the nascent field of synthetic biology and security.[7]

Lack of Accountability and Transparency

An outside observer of a multimillion-dollar federally funded center might well assume that there are regular bureaucratic procedures of audit, evaluation, and accountability in place. To a degree there are: after all, every spring at SynBERC there is a site review in which a team of reviewers informs the Center of their recommendations as to what the Center needs to do in the coming years to improve its performance. However, despite appearances, there was basically little or no contractual value in these reviews as in fact they did not actually provide the basis on which the next year's funding and evaluations would proceed.

Each year, we took the previous year's evaluation and recommendations as guides in formulating our work. Of course, we knew the situation was more complicated. We knew and were frequently reminded by PIs and SynBERC administrators alike that the Center's molecular biologists and engineers had significant multi-year funding from other sources. Consequently, they were unlikely to change course in how they cast their work unless there was a forceful reason—or threat—to their funding base beyond SynBERC. As we did not have another funding base, we felt obliged to adhere more literally to the formal ground rules. Such adherence came at the cost of an immense amount of bureaucratic labor that others in the Center neither felt obliged nor were required to undertake in anything like the detail demanded of us. In sum, we were constantly being given a double message: Be scientific leaders in your field, but present your work in a manner that conforms to the divisional leadership and IAB's limited and pre-formed expectations.

During the first two years, while insisting on attaining human scientific credibility in our work, we also committed ourselves to finding a mode of presentation through which the SynBERC PIs could at least understand why we were working on what we were working on rather than the first-order deliverables and affirmations of progress that they seemed to desire. After two years of laboring in this mode, and effectively sacrificing much valuable time by attending to the blockages and demands for clarity and justification on other people's terms (a demand no other thrust or PI was subjected to), we made the choice to pursue our work as much as possible in its integrity rather than to spend our time explaining ourselves to an audience that was fundamentally indifferent. In a sense, this approach was in line with what the other PIs at SynBERC were doing. They were more experienced in the arts of providing granting agencies what they needed in terms of boiler-plate prose and continuing apace with their own research as they saw fit. Our colleagues were "realists."

In year four, the IAB, through its leader, made it very clear that our mode of inquiry, which had passed high-level scientific standards of publication and peer review, did not interest or concern them: if they did not find it readily comprehensible and if it did not seem obviously instrumentally useful to their companies, then they would not count it as credible. For example, when we questioned the sufficiency of codes of conduct or elementary safety standards as adequate instruments of security for an emergent domain like synthetic biology, which promised massively increased capacities to manipulate DNA as well as a global distribution of these capacities through the Internet and the iGEM competitions, we were ignored. Demands for such codes and the like were repeated ultimately as threats to our funding. However, when such skeletal codes were produced (by our human practices colleagues at MIT), they received little or no response.

After five years of coexistence in the same organization, one of the senior SynBERC PIs asked Bennett, at an NIH workshop, how it felt to be SynBERC's "trophy wife." As vulgar as this phrasing was, it nonetheless resonated to the general situation in which we found ourselves.

Administrative Neglect

We began our experiment with the challenge—and an official mandate— to invent a post-ELSI mode of work. Our initial orientation was to shift from an ELSI mode of *cooperation* in which ethical, legal, and societal considerations were outside and downstream from the scientific and engineering work to a mode of *collaboration* in which common problems would be articulated between the multiple bioscientific and engineering disciplines represented in SynBERC and the human science disciplines we represented. At the end of this experiment, it is now clear that under the current conditions of significant power asymmetries, existing character structures both in the biosciences and the dominant human sciences, and pedagogical formation in the university system, collaboration almost certainly will not work without basic modifications. Reluctantly and regrettably, we have concluded that any mode of engagement that would focus ethics on a change in habits and dispositions—the guiding premise of upstream and midstream engagements—will be ignored or blocked. The existing scholarly literature substantiates the generality of this conclusion.[8] Our own networking with other centers working on parallel projects further confirms these conclusions.

We conclude that the Bush years of a politics and an ethos of little or no external regulation, of market forces as the dominant metric, of little or no separation of industry and science—all continue to reign in domains of promise and risk that one might well have assumed, if one was truly naive, that they had never fully penetrated. We have seen the disastrous consequences of this mode of insufficient governance in multiple other domains such as finance, food safety, environmental protection, and the like. In that light, we conclude that there is every reason to be concerned about the ramifications of developing areas of the life sciences such as synthetic biology.

SECTION II. LESSONS LEARNED

When all is said and done, we had made a bargain with SynBERC for which we take responsibility. We willingly engaged in an arrangement in which we received funding for our work and anticipated that our efforts to establish a Mode 3 form of engagement—efforts designed to eschew familiar first-order deliverables—would be supported in the spirit of scientific collegiality. From the outset, we probably should have known better, and on some level we did. After all, the lineage of twentieth-century thinkers whom we admire and have learned so much from—Max Weber, Michel Foucault, and John Dewey—all saw, in different ways and at different points in the last century (Weber at its beginning, Dewey halfway through, and Foucault toward its end), that the price to be paid for the power and instrumental mastery of modern science was the abandonment of hermeneutical meaning, general cultural significance, enhanced moral practice, and political or ethical spirituality. We underestimated the existential price to be paid. That being said, we did carry out a scientific experiment and in no way regret having done so.

Nietzsche had identified a similar crossroads in many of his works, especially in *Thus Spake Zarathustra*, where a mocking Zarathustra encounters "the last men who invented happiness," throws them a haughty glance, and continues his ascent to higher and purer realms. Weber ultimately counseled a Stoic disinterest offering little hope of breaking free, only the chilled, compensatory satisfaction of observation and analysis. Dewey, almost ever the optimist, after a half century in which his pragmatic hopes for bringing the sciences into a moral disposition were not so much thwarted as ignored, could only express his irritation, a sentiment he had no ethical concepts to elaborate.

Our philosophic and anthropological *daimon* has reminded us of these lessons as our experiment draws to an end. Max Weber's aside in "Science as a Vocation" remains pertinent: Might there not be a way to proceed once one understands that the natural sciences offer "no" answer?

Toward an Ethic of Flourishing

Today the dominant metrics of modernity—prosperity and amelioration—are entrenched and largely unexamined at least within the biosciences. In the last several decades, they have been institutionalized in bioethics and related disciplines as the reassuring outside to scientific research dedicated to the development of instrumentalities. Although they have been critically scrutinized by important bioethicists such as Leon Kass, these criticisms have not significantly altered the overall institutional practice and venues of the bioethics profession. In this apparatus, questions of ethics and the good are accounted for by a cooperative division of labor. To the extent that those on the human sciences side of the divide are willing to take up their work under the metrics of amelioration and prosperity—and many, if not all, are—synthetic biology and related areas of the post-genomic life sciences are certain to provide slots and funding for continued cooperation. The field of applicants to fill this slot is already crowded both in the United States and in Europe.

At the end of our experiment, we would add that the best form for increasing capacities is one that distributes tasks among and between individuals and organizations. In the human sciences in the twenty-first century, we are seeking a better form, venue, and set of practices than the still regnant valorization of virtuosity in which a single individual or organization is capable of performing all of the tasks necessary to diagnose, analyze, figure, and inquire into a particular problem-space arising within a broader problematization. Rather, we are convinced that our experiments and our experiences, here as elsewhere, point precisely to the need and necessity for collaboration.

Science as a Vocation: Inquiry and Vigilance

We are committed to a form of life that combines a reflective and rigorous combination of experience and experiment with forms of knowledge and ethics: *Human science is a vocation*. The vehicle through which that commitment is made pragmatic is inquiry: *Reason is a practice*. Inquiry entails an obligation to stay true to the demands necessary to give form to this practice and this vocation: *Thinking is risky*.

Inquiry requires discovering and formulating the conditioning elements of the problem at hand. Thus, inquiry can be said to be situated. Its goal is to isolate the vectors and interfaces in the world that are causing or occasioning indeterminacy or discordancy. Since the source and site of the problem lies in the situation (of which the observer is a part), it is only through discovering and giving tentative form to elements that are already present that the inquiry can precede. Hence the process involves staying in the midst of things of the world but transforming them in specific ways so as to give them the kind of form that is determinate and can be known, at least in principle. Form giving—often figural—is thus an essential goal and an essential moment of "describing" a problem and of shaping an inquiry.

Such a form of life demands a continuous, reiterative, and recursive process. Inquiry begins midstream, in a situation, both determinative and under-determined, and moves on (through the process of inquiry itself) to other situations and other problems, themselves both partially stabilized and partially troubled. Thus, it is perfectly appropriate—even rigorous—to begin with tentative parameters of the situation to be inquired into and tentative understandings of what is at stake. The question we pose at this stage in our experiment is this: What are the parameters and determinations that might help us design practices of inquiry under a metric of flourishing? The solution to a particular problem consists in a series of interventions whose particularities are not known before they are undertaken. The observation and reflection on the process can be called reason as long as we are clear that reason is neither a faculty of mind nor a quality of the things themselves, but rather a distinctive mode of taking up the practice of inquiry.

Maxims for Vigilance: Exercising *Paraskeue*

Given our vocational commitment to inquiry, the question remains: What is the ethic of such a practice? And given the blockages and difficulties we have encountered in our experiment with synthetic biology, how do we stay true to our vocation? We do not have a general answer, or a theory of ethics, to provide the reader. At this point, at the end of this experiment, however, we have gathered a set of maxims—*paraskeue*—whose deployment in everyday situations of discord and indeterminacy we have found to be invaluable as a means to restabilize our practice, to protect ourselves and our work from the vicissitudes, indifference, and violence of the world of knowledge-producers authorized to speak the truth about *anthropos* and *bios*, and to guide us in our experimental efforts.

Each of these maxims is an element that, when properly deployed, contributes to the maintenance of vigilance against the siren song calling for acquiescing to the reigning demands for instrumentality as the only end and only metric of scientific and ethical work. These maxims are meant to serve as *paraskeue* or equipment in daily life when one loses one's way, when attacks occur, or when demands, which one considers illegitimate, are accompanied by threats to one's livelihood and vocational integrity. Such threats are made on multiple scales and appear in multiple manners often unexpectedly, even in situations of familiarity that one might have assumed would contribute to a mutual trust and respect.

These maxims are not primarily designed as tools for the exercise of power. Rather, their essential function is as equipment of ethical vigilance. They are aids in keeping one on the scientific course to which one is committed. Naturally, a scientific course is not a road map of known territory. It is an exploration of uncertain terrain in which one is simultaneously exploring and giving form. Hence the incessant demand to deliver the substance and form of answers to scientific indeterminacies and ethical discordancies before one has conducted one's work constitutes a fundamental betrayal of science to which a loyalty to one's commitment to inquiry and a constant vigilance—armed with *paraskeue*—are the best defense. And since successful inquiry by definition opens up further problems, the demand for unambiguous deliverables that would stop this process in a definitive manner is itself equally a betrayal of science—and hence unethical.

We leave the reader with the following series of maxims (and glosses). These maxims are a distillation of our, and others', experiences. Throughout this experiment, the recourse to the exercise of these maxims has proven worthwhile, often leading to enhanced capacities to think and conduct our work. Most generally, their function is an aid in avoiding, enduring, and responding to the inevitable exercise of unequal power relations, many of whose ramifications have proved to be nefarious. These maxims are designed to fortify and encourage the resolute acceptance of the existence of ongoing risks of uncertainty and error that scientific practice entails.

As these maxims are quite general in form, in each experiment and experience they must be adjusted and taken up by practitioners in ways that suit their situation. In sum, we are not presenting a method here. Rather, these are signposts and tools for course correction, scientific animation, and a certain solace, and hopefully they will serve as points of orientation for work to come. If these (and other) maxims are to do their intended work, human scientists are challenged to make them their own.

1. Practice Problematization

Our form of inquiry is oriented to the near future. We seek to reformulate blockages and opportunities as problems. By so doing, one might be able to move toward making available a range of possible solutions; or at least to provide insight into the current shape and specificity of the problems at hand.

In our experiment, the question of what is being problematized is approached by identifying the ways in which formerly stable figures and their elements are becoming recombined and reconfigured. These figures should not be approached as epochal; they are not simply replacing prior figures. Rather, they should be approached in light of the fact that they share elements of existing figures in the process of recombination and reconfiguration. Thus, a primary task of problematization is to identify the relations among and between figures and their elements, and to identify vectors through which transformations are taking place and distinctive forms might be taking shape.

2. Remain Adjacent

Problematization is done best from a position of adjacency. Adjacent is defined as "situated near or close to something or each other, especially without touching." As opposed to traditional observation or participation, adjacency is explicitly allows inquiry to move in multiple directions. It thereby provides the potential for moving both into and out of an experimental situation in question as well as moving between different sites and scales. Such positionality requires constant calibration; therefore, tolerance for such frequent adjustment is an imperative for the conduct of this work.

3. Be Resolutely Second-Order

Adjacency works best as a second-order practice. The term *second-order* denotes a mode observation-intervention (*Betrachtungen*) in which the task, to use Niklas Luhmann's cryptic but incisive phrase, is to "observe observers observing."

Second-order practice proceeds through pedagogy and the vigilant assessment of events. It can be legitimately understood as a form of pedagogy both for oneself and others. Pedagogy involves reflective processes by which one builds the capacities for flourishing. Pedagogy is not equivalent to training, whose norm is expert knowledge. Rather, it involves the development of a disposition to learn how one's practices and experiences form or deform one's existence. In this instance, it is vigilant about how the sciences, understood in the broadest terms, enrich or impoverish those dispositions. Pedagogy teaches that flourishing is a lifelong formative and collaborative process.

A second set of concerns involves events that produce significant change in objects, relations, purposes, and modes of evaluation and action. By definition, these events cannot be adequately characterized until they begin to emerge. A core premise of our second-order participation is that in order to identify significant events, we need a sufficient tool kit of diagnostic and analytic concepts as well as a disposition of vigilant attention to the details of a situation as it unfolds.

Second-order practices are disruptive in that (at a minimum) they make visible existing habits and dispositions; this visibility often leads to the recognition that such dispositions and habits are insufficient and inadequate on one or another register. The identification of insufficiencies and inadequacies is frequently met by irritation, indifference, and the assertion of power to block or silence second-order observations.

As a defense and refusal, second-order practices are likely to provoke the demand for justification in the name of first-order stakes and deliverables. One must resist this demand. Inquiry conducted in a second-order mode is unlikely to produce outcomes that contribute directly to either the goals of prosperity or amelioration. This is a proverbial price worth paying.

4. Practice Frank Speech

Problematization, adjacency, and second-order observation, taken together, open a distinctive critical position that anchors and completes this series. Unless the insights produced are put into play in a serious and consequential manner, their ramifications will not be open to scrutiny. And their salutary effects on the practice of thinking will thereby be deflected or distorted.

Second-order observation aims to clarify habits and dispositions. It opens up the question as to what might need to be changed or maintained on the part of those observed as well as those observing. Such observation, as dispassionate or prosaic as it might be, is likely to provoke a reaction on the part of those observed. In our experience and the experience of others, this reaction, at least initially, oscillates between the poles of indifference and violence. One must remain alert to the fact that frank speech entails real dangers.

By speaking the truth frankly, however, the truth is made actual. Making truth claims in this mode allows for the possibility of unforeseen ramifications. Moreover, and more importantly, practicing frank speech in consequential situations actually makes one more capable of seeking the truth. This practice is thus scientifically and ethically worthwhile. This claim and its associated demands, taken up in an equipmental mode, are worth keeping ready at hand as one practices inquiry. By so doing, it makes a life of science a bit more risky, a bit more worthwhile, and, if one is fortunate, a bit more pleasurable.

NOTES

Introduction

1. John Dewey, introduction to *Essays in Experimental Logic* (1916; repr., New York: Dover, 1953), 11.
2. Ibid., 13.
3. Ian Hacking, "Historical Ontology," in *Historical Ontology* (Cambridge, MA: Harvard University Press, 2002), 1.
4. Dewey, *Experimental Logic*, 11.
5. Niklas Luhmann, *Observations on Modernity* (Palo Alto, CA: Stanford University Press, 1998).
6. A. Ganguli-Mitra et al., "Of Newtons and Heretics," *Nature Biotechnology* 27 (2009): 321–22.
7. The stakes of collaboration cohere with a maxim articulated by Michel Foucault: to increase capacities while not increasing relations of domination or exploitation. In the life sciences, this maxim can be rephrased as how to increase capacities without eliminating established understandings of the biology that, while currently troublesome, cannot simply be designed away. In the human sciences, this maxim stands as Foucault formulated it: however, given a fundamental inequality of power relations between the human and life sciences, which capacities have the potentiality to be increased and why they should be are the core challenges of diagnosis and inquiry. Successfully bringing these two disparate challenges together in a productive mode of collaboration is thus fraught, and how to do so remains unanswered and under-addressed.
8. Paul Rabinow and Tom White, "American Moderns," in *Essays on the Anthropology of Reason*, by Paul Rabinow (Princeton, NJ: Princeton University Press, 1996); Steve Shapin, *The Scientific Life: A Moral History of a Late Modern Vocation* (Chicago: University of Chicago Press, 2008).
9. A few of the more senior SynBERC scientists are aware that the security environment within which they are working must be addressed if their whole enterprise is to continue. Some understand that the stakes are significantly higher than the preservation of their own enterprise.
10. Sheila Jasanoff, *Designs on Nature: Science and Democracy in Europe and the United States* (Princeton, NJ: Princeton University Press, 2005); Steven Shapin, *The Scien-*

tific Life: A Moral History of a Late Modern Vocation (Chicago: University of Chicago Press, 2008).

11. See, for example, Drew Endy, "Foundations for Engineering Biology," *Nature* 438 (November 24, 2005): 449–53; David Baker et al., "Engineering Life: Building a FAB for Biology," *Scientific American* 294 (June 2006): 44–51.

12. See further http://bios-technika.net/.

Chapter One

1. Michel Foucault, "What Is Enlightenment?" in *The Foucault Reader*, ed. Paul Rabinow (New York: Pantheon Books, 1984), 48.

2. Daniel G. Gibson et al., "One-Step Assembly in Yeast of 25 Overlapping DNA Fragments to Form a Complete Synthetic *Mycoplasma genitalium* Genome," *Proceedings of the National Academy of Sciences USA* 105, no. 51 (December 23, 2008): 20404–9; David Baker et al., "Engineering Life: Building a FAB for Biology," *Scientific American* 294 (June 2006): 44–51; Paul Rabinow, *Anthropos Today: Reflections on Modern Equipment* (Princeton, NJ: Princeton University Press, 2002).

3. Sydney Brenner, "The End of the Beginning," *Science* 287 (2000): 2173–74.

4. The pillars that support genomic-sequencing models of organizing scientific research as well as the ethics of such research, while deteriorating, are still standing: first, the generation of interest and ample funding based on manifestos and their skillful dissemination; second, the construction of technology and appropriate facilities (including start-up companies); third, scientific advance that can be articulated with the manifestos; eventually, attention (or lip service) to ethical, legal, and social consequences.

5. On the "Code of Codes," see Daniel Kevles and Leroy Hood, *The Code of Codes* (Cambridge, MA: Harvard University Press, 1993). On one ambitious attempt to move directly from sequence information to diagnostic applications, see Paul Rabinow and Talia Dan Cohen, *A Machine to Make a Future: Biotech Chronicles* (Princeton, NJ: Princeton University Press, 2005).

6. On the genome-sequencing projects, the best books are Robert Cook-Deegan, *The Gene Wars: Science, Politics, and the Human Genome Project* (New York: Norton, 1995), and Robert Shreeve, *The Genome War: How Craig Venter Tried to Capture the Code of Life and Save the World* (New York: Knopf, 2004).

7. Baker et al., "Engineering Life"; Drew Endy, "Foundations for Engineering Biology," *Nature* 438 (November 24, 2005): 449–53.

8. Stephen Maurer and Laurie Zoloth, "Synthesizing Biosecurity," *Bulletin of the Atomic Scientists*, November 2007.

9. Paul Rabinow, "The Biological Modern," ARC Concept Note No. 6, February 2006, http://anthropos-lab.net/wp/publications/2007/08/conceptnote06.pdf (accessed December 1, 2010).

10. Although situated at this time as an anthropological observer in relation to SynBERC, Rabinow had been engaged with his colleagues Stephen Collier and Andrew

Lakoff in a project on the Global Biopolitics of Security that was interfacing with policy makers, even if that interface never developed into anything fruitful.

11. National Science Foundation ERC Year Three SynBERC Site Visit Team, "Third Year Renewal Report," February, 2009, 6–7.

12. In 2007 Jay Keasling and colleagues received funding for their proposed Joint Bio-Energy Institute (JBEI), http://jbei.org/.

13. NSF ERC Year Three, "Third Year Renewal Report," 19–20.

Chapter Two

1. This rhetoric still circulates. A summary statement of the significance of the Human Genome Project found on the U.S. National Genome Research Institute website reads, "Completed in April 2003, the HGP gave us the ability, for the first time, to read nature's complete genetic blueprint for building a human being." "All about the Human Genome Project (HGP)," http://www.genome.gov/10001772 (accessed June 15, 2011).

2. Robert Cook-Deegan, *The Gene Wars: Science, Politics, and the Human Genome Project* (New York: Norton, 1995).

3. On the concept of equipment, see Michel Foucault, *L'Herméneutique du sujet: Cours au Collège de France, 1981–82* (Paris: Hautes Etudes, Seuil/Gallimard, 2001); Foucault, *Sécurité, Territoire, Population: Cours au Collège de France, 1977–78* (Paris: Hautes Etudes, Seuil/Gallimard, 2004); and Paul Rabinow, *French Modern: Norms and Forms of the Social Environment* (Chicago: University of Chicago Press, 1989). For more on the technical meaning of "equipment" in ethics, see the Anthropological Research on the Contemporary at www.anthropos-lab.net.

4. See http://bioethics.georgetown.edu/pcbe/transcripts/jan02/index.html (accessed May 5, 2011).

5. See http://bioethics.georgetown.edu/pcbe/transcripts/jan02/jan17session.1.html.

6. http://conference.syntheticbiology.org (accessed October 1, 2006).

7. The phrase is from Philip Pauly, *Controlling Life: Jacques Loeb and the Engineering Ideal in Biology* (Oxford: Oxford University Press, 1987).

8. See http://ung.igem.org/Main_Page.

9. See the Sloan Report on synthetic biology, Michele S. Garfinkel et al., "Synthetic Genomics: Options for Governance" (October 2007), at http://www.jcvi.org/cms/fileadmin/site/research/projects/synthetic-genomics-report/synthetic-genomics-report.pdf, and the Fink Report on dual-use, CRS Report for Congress, *Oversight of Dual-Use Biological Research: The National Security Advisory Board for Biosecurity* (April 27, 2007), at http://www.fas.org/sgp/crs/natsec/RL33342.pdf.

10. Gaymon Bennett, "On the Care of Human Dignity" (PhD diss., Graduate Theological Union, 2008).

Chapter Three

1. See the Venter Institute's Carole Lartigue et al., "Genome Transplantation in Bacteria: Changing One Species to Another," *Science* 317, no. 5838 (August 3, 2007): 632–38.

2. See ProtoLife at http://www.protolife.net/.

3. See the BioBricks Foundation at http://biobricks.org/.

4. As we have argued in the introduction, standard *cooperative* models of science and society, such as those developed under the Human Genome Initiative's ELSI program, need to be adjusted and remediated. By "adjusted" we mean that given the significant changes that have taken place in the biosciences during the last decade, the core components of the ELSI program, developed to couple with the early stages of the genome-sequencing projects, today need recalibration. The success of such work depends on a number of factors, not least of which is the challenge of introducing new habits and forms of organization into the existing structures and practices of elite science.

5. Steven Shapin, *A Social History of Truth: Civility and Science in Seventeenth-Century England* (Chicago: University of Chicago Press, 1995); Andrew Barry, *Political Machines: Governing a Technological Society* (London: Athlone Press, 2001).

6. Niklas Luhmann, *Die Gesellschaft der Gesellschaft*, 2 vols. (Frankfurt-am-Main: Suhrkamp, 1997), 1:117.

7. Microsoft Word Encarta dictionary definition.

8. Max Weber, "Science as a Vocation," in *From Max Weber: Essays in Sociology*, trans. Hans Girth and C. Wright Mills (New York: Oxford University Press, 1941); Ulrich Beck, Anthony Giddens, and Scott Lash, *Reflexive Theory of Modernization* (London: Polity Press, 1994); Luc Boltanski and Laurent Thévenot, *On Justification: Economies of Worth*, trans. Catherine Porter (orig. 1991; Princeton, NJ: Princeton University Press, 2006).

9. Reinhart Koselleck, *Futures Past: On the Semantics of Historical Time*, trans. Keith Tribe (orig. 1979; Cambridge, MA: MIT Press, 1985).

10. Gregory Pence, *Who Is Afraid of Human Cloning?* (New York: Rowman and Littlefield, 1998).

11. Gilles Deleuze and Felix Guattari, *What Is Philosophy?*, trans. Hugh Tomlinson and Graham Burchell (orig. 1991; New York: Columbia University Press, 1994). In his book *L'éthique: Essai sur la conscience du Mal* (Paris: Hatier, 1995), Alain Badiou writes that one calls "opinions les représentations sans vérité, les débris anarchiques du savoir circulant. Or les opinions sont le ciment de la socialité. . . . L'opinion est la matière première de toute communication" (46).

12. Helta Nowotny, Peter Scott, and Michael Gibbons, *Re-thinking Science, Knowledge and the Public in an Age of Uncertainty* (London: Polity Press, 2001); Michael Gibbons et al., *The New Production of Knowledge: The Dynamics of Science and Research in Contemporary Societies* (New York: Sage, 1994).

13. See the BIOS Center, London School of Economics, www.lse.ac.uk/bios; Center for Bioethics and Medical Humanities, University of South Carolina, http://www.ipspr.sc.edu/cbmh/default.asp.

14. CNS-ASU homepage at http://cns.asu.edu/.

15. Erik Fisher, "Ethnographic Invention: Probing the Capacity of Laboratory Decisions," *NanoEthics* (July 2007), at http://www.cspo.org/documents/Fisher_Probing LabCapacity_Nanoethics-07.pdf.

16. http://cns.asu.edu/program/.

17. Ibid.

18. Ibid.

19. Marilyn Strathern, ed., *Audit Culture: Anthropological Studies in Accountability, Ethics, and the Academy* (London: Routledge, 2000).

20. Sheila Jasanoff, *Designs on Nature: Science and Democracy in Europe and the United States* (Princeton, NJ: Princeton University Press, 2005).

21. Paul Rabinow, *Marking Time: On the Anthropology of the Contemporary* (Princeton, NJ: Princeton University Press, 2007).

22. John Dewey, *Reconstruction in Philosophy*, enl. ed. (Boston: Beacon Press, 1957).

23. On the notion of a "History of the Present," see Michel Foucault, *Discipline and Punish*, trans. Alan Sheridan (orig. 1975; New York: Vintage Books, 1977).

Chapter Four

1. Encarta online dictionary.

2. Kenneth Burke, *Permanence and Change* (Indianapolis: Bobbs-Merrill, 1965), 97.

3. Max Weber, "Objectivity in Social Science and Social Policy," in *The Methodology of the Social Sciences*, trans. Edward Shils and Henry Finch (New York: Free Press, 1949).

4. Barry Canton, Anna Labno, and Drew Endy, "Refinement and Standardization of Synthetic Biological Parts and Devices," *Nature Biotechnology* 26 (2008): 787–93; Julius B. Lucks et al., "Toward Scalable Parts Families for Predictable Design of Biological Circuits," *Current Opinion in Microbiology* 11, no. 6 (December 2008): 567–73; Drew Endy, "Foundations for Engineering Biology," *Nature* 438 (November 24, 2005): 449–53.

5. As MIT's part's registry puts it: "Assembly of parts into devices and systems is being performed using traditional cloning techniques with a set of restriction sites that allow easy composition of composite devices that, in turn, can themselves be used as parts. Simultaneous parallel assembly lets us build many biological systems quickly." http://partsregistry.org/Main_Page (accessed May 5, 2011).

6. Adam Arkin, "Setting the Standard in Synthetic Biology," *Nature Biotechnology* 26 (2008): 771–74.

7. David Baker et al., "Engineering Life: Building a FAB for Biology," *Scientific American* 294 (June 2006): 44–51.

8. A. Rai and J. Boyle, "Synthetic Biology: Caught between Property Rights, the Public Domain, and the Commons," *PLoS Biology* 5, no. 3 (2007): 58; Hans Bügl et al., "DNA Synthesis and Biological Security," *Nature Biotechnology* 25, no. 6 (June 2007): 627–29.

9. A. Ganguli-Mitra et al., "Of Newtons and Heretics," *Nature Biotechnology* 27 (2009): 321–22.

10. M. O'Malley et al., "Knowledge-Making Distinctions in Synthetic Biology," *Bio-Essays* 30, no. 1 (January 2008): 57–65.

11. Dae-Kyun Ro et al., "Production of the Antimalarial Drug Precursor Artemisinic Acid in Engineered Yeast," *Nature* 440 (April 13, 2006): 940–43.

12. C. H. Martin et al., "Synthetic Metabolism: Engineering Biology at the Protein and Pathway Scales," *Chemistry and Biology* 16, no. 3 (March 27, 2009): 277–86.

13. Eric J. Steen et al., "Metabolic Engineering of *Saccharomyces cerevisiae* for the Production of n-butanol," *Microbial Cell Factories* 7 (2008): 36.

14. Joint BioEnergy Institute at http://jbei.org.

15. Yasuo Yoshikuni and Jay D. Keasling, "Pathway Engineering by Designed Divergent Evolution," *Current Opinion in Chemical Biology* 11, no. 2 (April 2007): 233–39.

16. Priscilla E. M. Purnick and Ron Weiss, "The Second Wave of Synthetic Biology: From Modules to Systems," *Nature Reviews Molecular Cell Biology* 10 (June 2009): 410–23; J. Ross and A. Arkin, "Complex Systems: From Chemistry to Systems Biology," *Proceedings of the National Academy of Sciences USA*, 106, no. 16 (April 21, 2009): 6433–34.

17. Daniel G. Gibson et al., "One-Step Assembly in Yeast of 25 Overlapping DNA Fragments to Form a Complete Synthetic *Mycoplasma genitalium* Genome," *Proceedings of the National Academy of Sciences USA* 105, no. 51 (December 23, 2008): 20404–9; T. Gabaldón et al., "The Core of a Minimal Gene Set: Insights from Natural Reduced Genomes," in *Protocells: Bridging Nonliving and Living Matter*, ed. Steen Rasmussen et al. (Cambridge, MA: MIT Press, 2008); Anthony C. Forster and George M. Church, "Towards Synthesis of a Minimal Cell," *Molecular Systems Biology* 2 (2006): article number 45, published online August 22, 2006, doi:10.1038/msb4100090v.

18. http://arep.med.harvard.edu/.

19. http://www.jcvi.org/.

20. Forster and Church, "Towards Synthesis of a Minimal Cell."

21. J. Tian et al., "Accurate Multiplex Gene Synthesis from Programmable DNA Microchips," *Nature* 432 (December 23/30, 2004): 1050–54.

22. Craig Venter, *A Life Decoded* (New York: Viking, 2007), 356.

23. Existing structures and processes can be either directly taken up or refashioned. Like Keasling, Venter wants to use organisms to produce specific molecules of interest. It is a step beyond redesigning pathways—redesigning genomes is an attempt to control all of the coding and reproduction operation.

24. Elise McCarthy and Christopher Kelty, "Responsibility and Nanotechnology," *Social Studies of Science* 40, no. 405 (2010): 1–28.

25. Michele Garfinkel et al., "Synthetic Genomics: Options for Governance," *Biosecurity and Bioterrorism: Biodefense Strategy, Practice, and Science* 4 (2007): 359–62; George Church, "Let Us Go Forth and Safely Multiply," *Nature* 438 (2005): 423.

26. Ganguli-Mitra et al., "Of Newtons and Heretics."

27. Purnick and Weiss, "The Second Wave"; Ross and Arkin, "Complex Systems."

28. P. Stano, G. Murtas, and P. L. Luisi, "Semi-Synthetic Minimal Cells: New Advancements and Perspectives," in *Protocells: Bridging Nonliving and Living Matter*, eds. S. Rasmussen et al. (Cambridge, MA: MIT Press, 2008).

29. Purnick and Weiss, "The Second Wave"; Lucks et al., "Toward Scalable Parts"; J. C. Anderson et al., "Environmentally Controlled Invasion of Cancer Cells by Engineered Bacteria," *Journal of Molecular Biology* 355 (2006): 619–27.

30. E. Andrianantoandro et al., "Synthetic Biology: New Engineering Rules for an Emerging Discipline," *Nature Molecular Systems Biology* 2 (2006): E1–14.

31. Paul Rabinow, *French DNA: Trouble in Purgatory* (Chicago: University of Chicago Press, 1999).

32. See Paul Berg, "Asilomar and Recombinant DNA," Nobelprize.org, http://nobelprize .org/nobel_prizes/chemistry/laureates/1980/berg-article.html (accessed June 21, 2011).

33. See Kenneth Burke, *Permanence and Change: An Anatomy of Purpose*, 3rd ed. (Berkeley: University of California Press, 1984), 7.

34. Norbert Elias, *The Court Society* (orig. 1969; Oxford: Basil Blackwell, 1983).

Chapter Five

1. For more on the notion of a biopolitical rationality and its relevance for synthetic biology, see our diagnostic work at http://bios-technika.net/.

2. Albert O. Hirschman, *Exit, Voice, and Loyalty: Responses to Decline in Firms, Organizations, and States* (Cambridge, MA: Harvard University Press, 1970).

3. National Science Foundation, "SynBERC, Fiscal Year 2008 Site Review Report."

Chapter Six

1. Max Weber, "Objectivity in Social Science and Social Policy," in *The Methodology of the Social Sciences*, trans. Edward Shils and Henry Finch (New York: Free Press, 1949), 68.

2. http://www.jcvi.org/.

3. http://www.thehastingscenter.org/Research/Detail.aspx?id=1548.

4. http://www.synbioproject.org/.

5. http://www.synbiosafe.eu/.

6. http://www.bbsrc.ac.uk/funding/opportunities/2007/synthetic_biology.html.

7. http://www.esf.org/activities/eurocores/programmes/eurosynbio.html.

8. Priscilla E. M. Purnick and Ron Weiss, "The Second Wave of Synthetic Biology: From Modules to Systems," *Nature Reviews Molecular Cell Biology* 10 (June 2009): 410–23.

9. Ideally, our table would also include policy work on intellectual property, which, thanks in large part to the efforts of Drew Endy and colleagues at MIT, has been a visible concern in synthetic biology from the outset. The question of how to reconcile an open-source ethos with industrial goals, however, involves a complicated and animated set of proposals that cannot be summarized here.

10. See, for example, the work of the National Scientific Advisory Board for Biosecurity, http://oba.od.nih.gov/biosecurity/biosecurity.html.

11. See, for example, Niklas Luhmann, *Risk: A Sociological Theory* (New York: Aldine Transaction, 2005).

12. http://syntheticbiology.org/SB2.0/Biosecurity_resolutions.html.

13. Stephen Maurer, Keith V. Lucas, and Starr Terrell, "From Understanding to Action: Community-Based Options for Improving Safety and Security in Synthetic Biology," 2006 whitepaper, http://gspp.berkeley.edu/iths/UC%20White%20Paper .pdf (accessed May 31, 2010).

14. Alexander Kelle, "Synthetic Biology and Biosecurity Awareness in Europe," *Bradford Science and Technology Report* no. 9 (2007).

15. Ibid.

16. A. Ganguli-Mitra et al., "Of Newtons and Heretics," *Nature Biotechnology* 27 (2009): 321–22.

17. Ibid.

18. Joachim Boldt and Oliver Müller, "Newtons of the Leaves of Grass," *Nature Biotechnology* 26 (2008): 387–89.

19. Helga Nowotny, Peter Scott, and Michael Gibbons, *Re-thinking Science: Knowledge and the Public in an Age of Uncertainty* (London: Polity Press, 2001).

20. Ganguli-Mitra et al., "Of Newtons and Heretics."

21. Laurie Zoloth, "Second Life: Some Ethical Issues in Synthetic Biology and the Recapitulation of Evolution," in *The Ethics of Protocells: Moral and Social Implications of Creating Life in the Laboratory*, ed. Mark Bedau and Emily Parke (Cambridge, MA: MIT Press, 2008); Erik Parens, Josephine Johnston, and Jacob Moses, "Do We Need 'Synthetic Bioethics'?" *Science* 321, no. 5895 (September 12, 2008): 1449; Boldt and Müller, "Newtons of the Leaves of Grass," 387.

22. Zoloth, "Second Life," 158–59.

23. Ibid., 160.

24. Ibid., 161.

25. Parens, Johnston, and Moses, "Do We Need 'Synthetic Bioethics'?," 1449.

26. In 1982 the National Commission on Biomedical and Biotechnical Research published the report "Splicing Life." The report represents a significant event in American bioethics in that it ratified and solidified the segregation of questions of safety and security—taken to be the purview of technical expertise—and questions of values and responsibility—the purview of bioethicists as representatives of "social" or "public" concern.

27. This conclusion is pressed against the findings of Boldt and Müller.

28. These "new scientific contexts" have been funded by the Sloan Foundation. Parens, Johnston, and Moses, "Do We Need 'Synthetic Bioethics'?," 1449.

29. Albert Jonsen, *The Birth of Bioethics* (Oxford: Oxford University Press, 1998).

30. Parens, Johnston, and Moses, "Do We Need 'Synthetic Bioethics'?," 1449.

31. Andrew Balmer and Paul Martin, "Synthetic Biology: Social and Ethical Challenges," Institute for Science and Society, University of Nottingham, May 2008, http://www .bbsrc.ac.uk/web/FILES/Reviews/0806_synthetic_biology.pdf.

32. M. O'Malley et al., "Knowledge-Making Distinctions in Synthetic Biology." *BioEssays* 30, no. 1 (January 2008): 57–65; George Khushf, "Upstream Ethics in Nano-

medicine: A Call for Research," *Nanomedicine* 2, no. 4 (September 2007): 511–21; Erik Fisher, Roop L. Mahajan, and Carl Mitcham, "Midstream Modulation of Technology: Governance from Within," *Bulletin of Science, Technology & Society* 26, no. 6 (December 2006): 485–96; Paul Rabinow, *Marking Time: On the Anthropology of the Contemporary* (Princeton, NJ: Princeton University Press, 2007); James Wilsdon and Rebecca Willis, *See-Through Science: Why Public Engagement Needs to Move Upstream* (London: Demos, 2004).

33. Jonsen, *The Birth of Bioethics*.

Chapter Seven

1. Priscilla E. M. Purnick and Ron Weiss, "The Second Wave of Synthetic Biology: From Modules to Systems," *Nature Reviews Molecular Cell Biology* 10 (June 2009): 410–23.
2. Ibid., 411.
3. Ibid., 412.
4. Barry Canton, Anna Labno, and Drew Endy, "Refinement and Standardization of Synthetic Biological Parts and Devices," *Nature Biotechnology* 26 (2008): 787.
5. Evelyn Fox Keller, "What Does Synthetic Biology Have to Do with Biology?" *BioSocieties* 4 (2009): 291–302.
6. Jason R. Kelly et al. "Measuring the Activity of BioBrick Promoters Using an *in Vivo* Reference Standard," *Journal of Biological Engineering* 3, no. 4 (2009).
7. Purnick and Weiss, "The Second Wave," 410.
8. Ibid., 412.
9. Ibid.
10. Ibid.
11. Ibid., 413.
12. Ibid.
13. Ibid., 414.
14. Ibid., 415.
15. Ibid., 414.
16. See, for example, Dan Gibson et al. "Complete Chemical Synthesis, Assembly, and Cloning of a Mycoplasma genitalium Genome," *Science* 319 (2008): 1215–20; and A. C. Forster and G. M. Church, "Towards Synthesis of a Minimal Cell," *Molecular and Systems Biology* 2 (2006): 45.
17. Purnick and Weiss, "The Second Wave," 415.
18. Ibid.
19. For the complete set of rules comprising the standard, see the OpenWetWare BioBrick standard web page at http://openwetware.org/wiki/Biobrick_standard (accessed May 3, 2011).
20. Purnick and Weiss, "The Second Wave," 416.
21. Ibid., 417.
22. Ibid.
23. Ibid.

24. Ibid.
25. Ibid., 418.
26. Ibid.
27. See Ali Zarrinpar, Sang-Hyun Park, and Wendell A. Lim, "Optimization of Specificity in a Cellular Protein Interaction Network by Negative Selection," *Nature* 426 (December 11, 2003): 676–80.
28. Max Weber, "Objectivity in Social Science and Social Policy," in *The Methodology of the Social Sciences*, trans. Edward Shils and Henry Finch (New York: Free Press, 1949), 68.

Chapter Eight

1. François Flahault, *La méchanceté* (Paris: La Decouverte, 1998). English translation by Liz Heron, *Malice* (London: Verso Books, 2003).
2. See Aristotle's *Posterior Analytics* and Girard Genette's *Narrative Discourse Revisited*, trans. Jane E. Lewin (Ithaca, NY: Cornell University Press, 1988).
3. We are aware of the hermeneutic controversies attached to Auerbach's work, to which we do not intend to enter. Stripped of this controversy, we find a central point that Auerbach makes extremely helpful to our work.
4. Although each of the figures consists of elements that themselves are characterized by other temporalities. Unlike Auerbach's figural interpretations, the ontological mode of the figures taken up in this diagnostic of the current biosecurity situation is characterized neither by the eternal nor transcendental nor historically comprehensive. Rather, the temporality that characterizes the ontological mode is *contemporary*.
5. Paul Rabinow and Nikolas Rose, "Biopower Today," *BioSocieties* 1 (2006): 195–217.
6. Ian Hacking, *Representing and Intervening: Introductory Topics in the Philosophy of Natural Science* (Cambridge: Cambridge University Press, 1983).
7. Robert Cook-Deegan, *The Gene Wars: Science, Politics, and the Human Genome Project* (New York: Norton, 1995); and Robert Shreeve, *The Genome War: How Craig Venter Tried to Capture the Code of Life and Save the World* (New York: Knopf, 2004).
8. Daniel Kevles and Leroy Hood, *The Code of Codes* (Cambridge, MA: Harvard University Press, 1993).
9. Paul Rabinow, *Marking Time: On the Anthropology of the Contemporary* (Princeton, NJ: Princeton University Press, 2007).
10. Ken Alibek, *Biohazard: The Chilling True Story of the Largest Covert Biological Weapons Program in the World—Told from Inside by the Man Who Ran It* (New York: Random House, 1999).
11. Flahault, *La méchanceté*.
12. Ibid., 2.
13. Ibid.
14. Ibid., 4.
15. Ibid., 165.

16. Niklas Luhmann, *Observations on Modernity* (Palo Alto, CA: Stanford University Press, 1998).

17. National Research Council, *Biotechnology Research in an Age of Terrorism: Confronting the Dual Use Dilemma* (Washington, DC: National Academies Press, 2004).

18. Ibid., viii.

19. Ibid., vii.

20. National Scientific Advisory Board for Biosecurity's "Frequently Asked Questions," http://oba.od.nih.gov/oba/ibc/FAQs/FAQs%20about%20the%20National%20Science%20Advisory%20Board%20for%20Biosecurity.pdf (accessed December 1, 2010).

21. Michele Garfinkel et al., "Synthetic Genomics: Options for Governance." *Biosecurity and Bioterrorism: Biodefense Strategy, Practice, and Science* 4 (2007): 359–62.

22. Elise McCarthy and Christopher Kelty, "Responsibility and Nanotechnology," *Social Studies of Science* 40, no. 405 (2010): 1–28.

23. Ibid.

24. http://www.ia-sb.eu/go/synthetic-biology; http://www.ia-sb.eu/tasks/sites/synthetic-biology/assets/File/pdf/iasb_code_of_conduct_final.pdf.

25. Michael Power, *Organized Uncertainty: Designing a World of Risk Management* (Oxford: Oxford University Press, 2007).

26. Flahault, *La méchanceté*, 4.

27. Ibid., 165.

28. Michel Foucault, Paul Veyne, and François Weil, "Des travaux," in *Dits et Ecrits, T. IV*, by Michel Foucault (Paris: Gallimard, 1983), 336.

Chapter Nine

1. Michel Foucault, Paul Veyne, and François Weil, "Des travaux" in *Dits et Ecrits, T. IV*, by Michel Foucault (Paris: Gallimard, 1983), 336.

2. François Flahault, *La méchanceté* (Paris: La Decouverte, 1998), 4. English translation by Liz Heron, *Malice* (London: Verso Books, 2003), 4.

3. Flahault, *La méchanceté*, 165.

4. Michael Power, *Organized Uncertainty: Designing a World of Risk Management* (Oxford: Oxford University Press, 2007).

5. Michel Foucault, *Sécurité, Territoire, Population: Cours au Collège de France, 1977–78* (Paris: Hautes Etudes, Seuil/Gallimard, 2004).

6. For more on this distinction, see Paul Rabinow, "Epilogue," in *French DNA: Trouble in Purgatory* (Chicago: University of Chicago Press, 1999).

7. Foucault, *Securité*, 13. This example is discussed at more length in Paul Rabinow, *French Modern: Norms and Forms of the Social Environment* (Chicago: University of Chicago Press, 1989).

8. Foucault, *Securité*, 21.

9. Ibid., 23.

10. Ibid., 48.

11. Niklas Luhmann, *Risk: A Sociological Theory* (New York: Aldine Transaction, 2005).

12. Ian Hacking, *The Emergence of Probability: A Philosophical Study of Early Ideas about Probability, Induction and Statistical Inference* (Cambridge: Cambridge University Press, 2006); Alain Desrosieres, *The Politics of Large Numbers: A History of Statistical Reasoning* (Cambridge, MA: Harvard University Press, 1998).

13. Recall that normalization designates a project to order aspects of things according to regular distributions. Norms constitute the grounds for normalization. If the elements picked out as a danger and as a risk are social, then the project of normalization is also normative. Values are assigned to different distributions of risk, and social interventions can be designed to affect these different distributions.

14. See http://www.synbiosafe.eu/.

15. Stasis: a state in which there is neither motion nor development, often resulting from opposing forces balancing each other. Greek thinkers, like Thucydides, used the term *stasis* to refer to a state of incipient or simmering warfare or hostilities.

16. John Dewey, *Reconstruction in Philosophy* (1920; new ed., Boston: Beacon Press, 1948), xxvii.

17. Rabinow, *French Modern*.

18. Priscilla E. M. Purnick and Ron Weiss, "The Second Wave of Synthetic Biology: From Modules to Systems," *Nature Reviews Molecular Cell Biology* 10 (June 2009): 412.

Chapter Ten

1. Although Rabinow was displaced as the head of Human Practices, he continued as a PI before resigning voluntarily from SynBERC in June 2011.

2. National Science Foundation ERC Year SynBERC Three Site Visit Team, "Fourth Annual Renewal Report," March 2010, 10–11.

3. National Science Foundation ERC Year SynBERC Three Site Visit Team, "Third Year Renewal Report," February 2009, 20.

4. Alexander Kelle, "Synthetic Biology and Biosecurity Awareness in Europe," *Bradford Science and Technology Report*, no. 9 (2007).

5. A. Ganguli-Mitra et al., "Of Newtons and Heretics," *Nature Biotechnology* 27 (2009): 321–22.

6. See, for example, Andrew Lakoff and Stephen J. Collier, eds., *Biosecurity Interventions: Global Health and Security in Question* (New York: Columbia University Press, 2008).

7. Gaymon Bennett, Nils Gilman, Anthony Stavriankis, and Paul Rabinow, "From Synthetic Biology to Bio-Hacking: Are We Prepared?" *Nature Biotechnology* 27 (2009): 1109–11; Paul Rabinow and Gaymon Bennett, "Synthetic Biology: Ethical Ramifications 2009," *Journal of Systems and Synthetic Biology* 3, nos. 1–4 (December 2009).

8. Kelle, "Synthetic Biology."

BIBLIOGRAPHY

Alibek, Ken. *Biohazard: The Chilling True Story of the Largest Covert Biological Weapons Program in the World—Told from Inside by the Man Who Ran It*. New York: Random House, 1999.

Anderson, J. C., E. J. Clarke, A. P. Arkin, and C. A. Voigt. "Environmentally Controlled Invasion of Cancer Cells by Engineered Bacteria." *Journal of Molecular Biology* 355 (2006): 619–27.

Andrianantoandro, E., S. Basu, D. K. Karig, and R. Weiss. "Synthetic Biology: New Engineering Rules for an Emerging Discipline." *Nature Molecular Systems Biology* 2 (2006): E1–14.

Arkin, Adam P. "Setting the Standard in Synthetic Biology." *Nature Biotechnology* 26 (2008): 771–74.

Badiou, Alain. *L'éthique: Essai sur la conscience du Mal*. Paris: Hatier, 1995.

Baker, David, George Church, Jim Collins, Drew Endy, Joseph Jacobson, Jay Keasling, Paul Modrich, Christina Smolke, and Ron Weiss. "Engineering Life: Building a FAB for Biology." *Scientific American* 294 (June 2006): 44–51.

Balmer, Andrew, and Paul Martin, "Synthetic Biology: Social and Ethical Challenges." Institute for Science and Society, University of Nottingham. May 2008. http://www .bbsrc.ac.uk/web/FILES/Reviews/0806_synthetic_biology.pdf.

Barben, Daniel, Erik Fisher, Cynthia Selin, and David Guston. "Anticipatory Governance of Nanotechnology: Foresight, Engagement, and Integration." In *The Handbook of Science and Technology Studies*, 3rd ed., edited by Edward J. Hackett, Olga Amsterdamska, Michael Lynch, and Judy Wajcman. Cambridge, MA: MIT Press, 2007.

Barry, Andrew. *Political Machines: Governing a Technological Society*. London: Athlone Press, 2001.

Beck, Ulrich, Anthony Giddens, and Scott Lash. *Reflexive Theory of Modernization*. London: Polity Press, 1994.

Bedau, Mark, and Emily Parke, eds. *The Ethics of Protocells: Moral and Social Implications of Creating Life in the Laboratory*. Cambridge, MA: MIT Press, 2008.

Bennett, Gaymon. "On the Care of Human Dignity." PhD diss., Graduate Theological Union, 2008.

Bennett, Gaymon, Nils Gilman, Anthony Stavrianakis, and Paul Rabinow. "From Syn-
    thetic Biology to Bio-Hacking: Are We Prepared?" *Nature Biotechnology* 27 (2009):
    1109–11.

Berg, Paul. "Asilomar and Recombinant DNA." Nobelprize.org. http://nobelprize.org/
    nobel_prizes/chemistry/laureates/1980/berg-article.html (accessed June 21, 2011).

Boldt, Joachim, and Oliver Müller. "Newtons of the Leaves of Grass." *Nature Biotechnol-
    ogy* 26 (2008): 387–89.

———. "Of Newtons and Heretics." *Nature Biotechnology* 27 (2009): 321–22.

Boltanski, Luc, and Laurent Thévenot. *On Justification: Economies of Worth*. 1991. Trans-
    lated by Catherine Porter. Princeton, NJ: Princeton University Press, 2006.

Brenner, Sydney. "The End of the Beginning." *Science* 287 (2000): 2173–74.

Bügl, Hans, et al. "DNA Synthesis and Biological Security." *Nature Biotechnology* 25,
    no. 6 (June 2007): 627–29.

Burke, Kenneth. *Permanence and Change*. Indianapolis: Bobbs-Merrill, 1965.

———. *Permanence and Change: An Anatomy of Purpose*. 3rd ed. Berkeley: University of
    California Press, 1984.

Canton, Barry, Anna Labno, and Drew Endy. "Refinement and Standardization of Syn-
    thetic Biological Parts and Devices." *Nature Biotechnology* 26 (2008): 787–93.

Church, George. "Let Us Go Forth and Safely Multiply." *Nature* 438 (2005): 423.

Cook-Deegan, Robert. *The Gene Wars: Science, Politics, and the Human Genome Project*.
    New York: Norton, 1995.

CRS Report for Congress. *Oversight of Dual-Use Biological Research: The National Se-
    curity Advisory Board for Biosecurity*. April 27, 2007. http://www.fas.org/sgp/crs/
    natsec/RL33342.pdf.

Daston, Lorraine, and Peter Galison. *Objectivity*. Cambridge, MA: MIT Press, 2007.

Deleuze, Gilles, and Felix Guattari. *What Is Philosophy?* 1991. Translated by Hugh Tom-
    linson and Graham Burchell. New York: Columbia University Press, 1994.

Desrosieres, Alain. *The Politics of Large Numbers: A History of Statistical Reasoning*.
    Cambridge, MA: Harvard University Press, 1998.

Dewey, John. *Essays in Experimental Logic*. 1916. Reprint, New York: Dover, 1953.

———. *Reconstruction in Philosophy*. 1920. New ed. Boston: Beacon Press, 1948.

———. *Reconstruction in Philosophy*, enl. ed. Boston: Beacon Press, 1957.

Elias, Norbert. *The Court Society*. 1969. Oxford: Basil Blackwell, 1983.

Endy, Drew. "Foundations for Engineering Biology." *Nature* 438 (November 24, 2005):
    449–53.

Evans, John. *Playing God?: Human Genetic Engineering and the Rationalization of Public
    Bioethical Debate*. Chicago: University of Chicago Press, 2001.

Fisher, Erik. "Ethnographic Invention: Probing the Capacity of Laboratory Decisions,"
    Nanoethics (2006), http://www.cspo.org/documents/Fisher_ProbingLabCapacity
    _Nanoethics-07.pdf (accessed July 13, 2011).

Fisher, Erik, Roop L. Mahajan, and Carl Mitcham. "Midstream Modulation of Tech-
    nology: Governance from Within." *Bulletin of Science, Technology & Society* 26, no. 6
    (December 2006): 485–96.

Flahault, François. *La méchanceté*. Paris: La Découverte, 1998. (English translation by Liz Heron. *Malice*. London: Verso Books, 2003.)

Forster, Anthony C., and George M. Church. "Towards Synthesis of a Minimal Cell." *Molecular Systems Biology* 2 (2006): 45.

Foucault, Michel. *Discipline and Punish*. Translated by Alan Sheridan. New York: Vintage Books, 1977.

———. *L'Herméneutique du sujet: Cours au Collège de France, 1981–82*. Paris: Hautes Etudes, Seuil/Gallimard, 2001.

———. *Sécurité, Territoire, Population: Cours au Collège de France, 1977–78*. Paris: Hautes Etudes, Seuil/Gallimard, 2004.

———. "What Is Enlightenment?" In *The Foucault Reader*, edited by Paul Rabinow. New York: Pantheon Books, 1984.

Foucault, Michel, Paul Veyne, and François Weil. "Des travaux." In *Dits et Ecrits, T. IV*, by Michel Foucault. Paris: Gallimard, 1983.

Gabaldón, T., R. Gil, J. Peteró, A. Latorre, and A. Moya. "The Core of a Minimal Gene Set: Insights from Natural Reduced Genomes." In *Protocells: Bridging Nonliving and Living Matter*, edited by Steen Rasmussen, Mark A. Bedau, Liaohai Chen, David Deamer, David C. Krakauer, Norman H. Packard, and Peter F. Stadler. Cambridge, MA: MIT Press, 2008.

Galison, Peter. *How Experiments End*. Chicago: University of Chicago Press, 1987.

Ganguli-Mitra, A., M. Schmidt, H. Torgersen, A. Deplazes, and N. Biller-Andorno. "Of Newtons and Heretics." *Nature Biotechnology* 27 (2009): 321–22.

Garfinkel, M., D. Endy, G. Epstein, and R. Freidman. "Synthetic Genomics: Options for Governance." *Biosecurity and Bioterrorism: Biodefense Strategy, Practice, and Science* 4 (2007): 359–62.

Genette, Girard. *Narrative Discourse Revisited*. Translated by Jane E. Lewin. Ithaca, NY: Cornell University Press, 1988.

Gibbons, Michael. "Science's New Social Contract with Society." *Nature* 402 (1999): C81.

Gibbons, Michael, Camille Limoges, Helga Nowotny, Simon Schwartzman, Peter Scott, and Martin Trow. *The New Production of Knowledge: The Dynamics of Science and Research in Contemporary Societies*. New York: Sage, 1994.

Gibson, Daniel G., Gwynedd A. Benders, Kevin C. Axelrod, Jayshree Zaveri, Mikkel A. Algire, Monzia Moodie, Michael G. Montague, J. Craig Venter, Hamilton O. Smith, and Cyde A. Hutchinson III. "One-Step Assembly in Yeast of 25 Overlapping DNA Fragments to Form a Complete Synthetic *Mycoplasma genitalium* Genome." *Proceedings of the National Academy of Sciences USA* 105, no. 51 (December 23, 2008): 20404–9.

Guston, David. *Between Politics and Science: Assuring the Integrity and Productivity of Research*. New York: Cambridge University Press, 2000.

Hacking, Ian. *The Emergence of Probability: A Philosophical Study of Early Ideas about Probability, Induction and Statistical Inference*. Cambridge: Cambridge University Press, 2006.

———. *Historical Ontology*. Cambridge, MA: Harvard University Press, 2002.

———. *Representing and Intervening: Introductory Topics in the Philosophy of Natural Science*. Cambridge: Cambridge University Press, 1983.

Hayden, Cori. *When Nature Goes Public: The Making and Unmaking of Bioprospecting in Mexico*. Princeton, NJ: Princeton University Press, 2003.

Hirschman, Albert O. *Exit, Voice, and Loyalty: Responses to Decline in Firms, Organizations, and States*. Cambridge, MA: Harvard University Press, 1970.

Jasanoff, Sheila. *Designs on Nature: Science and Democracy in Europe and the United States*. Princeton, NJ: Princeton University Press, 2005.

Jonsen, Albert. *The Birth of Bioethics*. Oxford: Oxford University Press, 1998.

Kelle, Alexander. "Synthetic Biology and Biosecurity Awareness in Europe." *Bradford Science and Technology Report*, no. 9 (2007).

Keller, Evelyn Fox. *The Century of the Gene*. Cambridge, MA: Harvard University Press, 2000.

———. *A Feeling for the Organism*. New York: Macmillan, 1984.

———. "What Does Synthetic Biology Have to Do with Biology?" *BioSocieties* 4 (2009): 291–302.

Kelly, Jason R., Adam J. Rubin, Joseph H. Davis, Caroline M. Ajo-Franklin, John Cumbers, Michael J. Czar, Kim de Mora, Aaron L. Glieberman, Dileep D. Monie, and Drew Endy. "Measuring the Activity of BioBrick Promoters Using an *in Vivo* Reference Standard." *Journal of Biological Engineering* 3, no. 4 (2009).

Kevles, Daniel, and Leroy Hood. *The Code of Codes*. Cambridge, MA: Harvard University Press, 1993.

Khushf, George. "Upstream Ethics in Nanomedicine: A Call for Research." *Nanomedicine* 2, no. 4 (September 2007): 511–21.

Kleinman, Arthur, Renee Fox, and Allan Brandt, eds. "Bioethics & Beyond." Special issue, *Daedalus* 128, no. 4 (1999).

Koselleck, Reinhart. *Futures Past: On the Semantics of Historical Time*. 1979. Translated by Keith Tribe. Cambridge, MA: MIT Press, 1985.

Lakoff, Andrew, and Stephen J. Collier, eds. *Biosecurity Interventions: Global Health and Security in Question*. New York: Columbia University Press, 2008.

Lartigue, Carole, John I. Glass, Nina Alperovich, Rembert Pieper, Prashanth P. Parmar, Clyde A. Hutchinson III, Hamilton O. Smith, and J. Craig Venter. "Genome Transplantation in Bacteria: Changing One Species to Another." *Science* 317, no. 5838 (August 3, 2007): 632–38.

Lash, Scott, Bronislaw Szerszynski, and Brian Wynne, eds. *Risk, Environment and Modernity: Towards a New Ecology*. London: Sage, 1996.

Latour, Bruno, and Stephen Woolgar. *Laboratory Life: The Social Construction of Scientific Facts*. Beverly Hills: Sage, 1979.

Lucks, Julius B., Lei Qi, Weston R. Whitaker, and Adam P. Arkin. "Toward Scalable Parts Families for Predictable Design of Biological Circuits." *Current Opinion in Microbiology* 11, no. 6 (December 2008): 567–73.

Luhmann, Niklas. *Die Gesellschaft der Gesellschaft*. 2 vols. Frankfurt-am-Main: Suhrkamp, 1997.

———. *Observations on Modernity*. Palo Alto, CA: Stanford University Press, 1998.

————. *Risk: A Sociological Theory*. New York: Aldine Transaction, 2005.

Martin, C. H., D. R. Nielsen, K. V. Solomon, and K. L. Prather. "Synthetic Metabolism: Engineering Biology at the Protein and Pathway Scales." *Chemistry and Biology* 16, no. 3 (March 27, 2009): 277–86.

Martin, Vincent J., Douglas J. Pitera, Sydnor T. Withers, Jack D. Newman, and Jay D. Keasling. "Engineering a Mevalonate Pathway in *Escherichia coli* for Production of Terpenoids." *Nature Biotechnology* 21 (2003): 796–802.

Maurer, Stephen, Keith V. Lucas, and Starr Terrell. "From Understanding to Action: Community-Based Options for Improving Safety and Security in Synthetic Biology." 2006. Whitepaper. http://gspp.berkeley.edu/iths/UC%20White%20Paper.pdf (accessed May 31, 2010).

Maurer, Stephen, and Laurie Zoloth. "Synthesizing Biosecurity." *Bulletin of the Atomic Scientists*, November 2007.

McCarthy, Elise, and Christopher Kelty. "Responsibility and Nanotechnology." *Social Studies of Science* 40, no. 405 (2010): 1–28.

National Research Council. *Biotechnology Research in an Age of Terrorism: Confronting the Dual Use Dilemma*. Washington, DC: National Academies Press, 2004.

National Science Foundation. "SynBERC: Fiscal Year 2008 Site Review Report."

National Science Foundation ERC Year SynBERC Three Site Visit Team. "Fourth Annual Renewal Report." March 2010.

————. "Third Year Renewal Report." February 2009.

Nowotny, Helga, Peter Scott, and Michael Gibbon. *Re-thinking Science: Knowledge and the Public in an Age of Uncertainty*. London: Polity Press, 2001.

O'Malley, M., A. Powell, J. F. Davies, and J. Calvert. "Knowledge-Making Distinctions in Synthetic Biology." *BioEssays* 30, no. 1 (January 2008): 57–65.

Parens, Erik, Josephine Johnston, and Jacob Moses. "Do We Need 'Synthetic Bioethics'?" *Science* 321, no. 5895 (September 12, 2008): 1449.

Pauly, Phillip. *Controlling Life: Jacques Loeb and the Engineering Ideal in Biology*. Oxford: Oxford University Press, 1987.

Pence, Gregory. *Who Is Afraid of Human Cloning?* New York: Rowman and Littlefield, 1998.

Power, Michael. *Organized Uncertainty: Designing a World of Risk Management*. Oxford: Oxford University Press, 2007.

Purnick, Priscilla E. M., and Ron Weiss. "The Second Wave of Synthetic Biology: From Modules to Systems." *Nature Reviews Molecular Cell Biology* 10 (June 2009): 410–22.

Rabinow, Paul. *Anthropos Today: Reflections on Modern Equipment*. Princeton, NJ: Princeton University Press, 2002.

————. "The Biological Modern." ARC Concept Note No. 6. February 2006. http://anthropos-lab.net/wp/publications/2007/08/conceptnoteno6.pdf (accessed December 1, 2010).

————. *French DNA: Trouble in Purgatory*. Chicago: University of Chicago Press, 1999.

————. *French Modern: Norms and Forms of the Social Environment*. Chicago: University of Chicago Press, 1989.

———. *Marking Time: On the Anthropology of the Contemporary*. Princeton, NJ: Princeton University Press, 2007.

———. *Symbolic Domination: Cultural Form and Historical Change in Morocco*. Chicago: University of Chicago Press, 1975.

Rabinow, Paul, and Gaymon Bennett. "From Bioethics to Human Practices, or Assembling Contemporary Equipment." In *Tactical Biopolitics Art, Activism, and Technoscience*, edited by Beatriz da Costa and Kavita Philips. Cambridge, MA: MIT Press, 2007.

———. "Human Practices: Interfacing Three Modes of Collaboration." In *The Ethics of Protocells: Moral and Social Implications of Creating Life in the Laboratory*, edited by Mark Bedau and Emily Parke. Cambridge, MA: MIT Press, 2008.

———. "Synthetic Biology: Ethical Ramifications 2009." *Journal of Systems and Synthetic Biology* 3, nos. 1–4 (December 2009).

Rabinow, Paul, and Talia Dan Cohen. *A Machine to Make a Future: Biotech Chronicles*. Princeton, NJ: Princeton University Press, 2005.

Rabinow, Paul, and Nikolas Rose. "Biopower Today." *BioSocieties* 1 (2006): 195–217.

Rabinow, Paul and Tom White. "American Moderns." In *Essays on the Anthropology of Reason*, by Paul Rabinow. Princeton, NJ: Princeton University Press, 1996.

Rai, A., and J. Boyle. "Synthetic Biology: Caught between Property Rights, the Public Domain, and the Commons." *PLoS Biology* 5, no. 3 (2007): 58.

Rasmussen, Steen, Mark A. Bedau, Liaohai Chen, David Deamer, David C. Krakauer, Norman H. Packard, and Peter F. Stadler, eds. *Protocells: Bridging Nonliving and Living Matter*. Cambridge, MA: MIT Press, 2008.

Rip, Arie, Thomas J. Misa, and Johan Schot, eds. *Managing Technology in Society*. London: Pinter, 1995.

Ro, Dae-Kyun, et al. "Production of the Antimalarial Drug Precursor Artemisinic Acid in Engineered Yeast." *Nature* 440 (April 13, 2006): 940–43.

Rose, Nikolas, et al. "Genomics." Special issue, *BioSocieties* 1, no. 1 (2006).

Ross, J., and A. Arkin. "Complex Systems: From Chemistry to Systems Biology." *Proceedings of the National Academy of Sciences USA* 106, no. 16 (April 21, 2009): 6433–34.

Schmidt M., A. Kelle, A. Ganguli, and H. de Vriend, eds. *Synthetic Biology: The Technoscience and Its Societal Consequences*. Berlin: Springer, 2009.

Shapin, Steven. *The Scientific Life: A Moral History of a Late Modern Vocation*. Chicago: University of Chicago Press, 2008.

———. *A Social History of Truth: Civility and Science in Seventeenth-Century England*. Chicago: University of Chicago Press, 1995.

Shreeve, Robert. *The Genome War: How Craig Venter Tried to Capture the Code of Life and Save the World*. New York: Knopf, 2005.

Snow, C. P. *The Two Cultures*. Cambridge: Cambridge University Press, 1960.

Stano, P., G. Murtas, and P. L. Luisi. "Semi-Synthetic Minimal Cells: New Advancements and Perspectives. In *Protocells: Bridging Nonliving and Living Matter*, edited by Steen Rasmussen, Mark A. Bedau, Liaohai Chen, David Deamer, David C. Krakauer, Norman H. Packard, and Peter F. Stadler. Cambridge, MA: MIT Press, 2008.

Steen, Eric J., Rossana Chan, Nilu Prasad, Samuel Myers, Christopher J. Petzold, Alyssa Redding, Mario Ouellet, and Jay D. Keasling. "Metabolic Engineering of *Saccharomyces cerevisiae* for the Production of n-butanol." *Microbial Cell Factories* 7 (2008): 36.

Strathern, Marilyn, ed. *Audit Culture: Anthropological Studies in Accountability, Ethics, and the Academy.* London: Routledge, 2000.

Tian, J., H. Gong, N. Sheng, X. Zhou, E. Gulari, X. Gao, and G. Church. "Accurate Multiplex Gene Synthesis from Programmable DNA Microchips." *Nature* 432 (December 23/30, 2004): 1050–54.

U.S. National Genome Research Institute. "All about the Human Genome Project." http://www.genome.gov/10001772.

Venter, Craig. *A Life Decoded.* New York: Viking, 2007.

Weber, Max. "Objectivity in Social Science and Social Policy." In *The Methodology of the Social Sciences.* Translated by Edward Shils and Henry Finch. New York: Free Press, 1949.

———. "Science as a Vocation." In *From Max Weber: Essays in Sociology*, edited by H. H. Gerth and C. Wright Mills. Oxford: Oxford University Press, 1958.

Wilsdon, James, and Rebecca Willis. *See-Through Science: Why Public Engagement Needs to Move Upstream.* London: Demos, 2004.

Yearly, Steven. "The Ethical Landscape: Identifying the Right Way to Think about the Ethical and Societal Aspects of Synthetic Biology Research and Products." *Interface: Journal of the Royal Academy.* May 15, 2009. http://rsif.royalsocietypublishing.org/content/early/2009/05/12/rsif.2009.0055.focus.short (accessed June 24, 2011).

Yoshikuni, Yasuo, and Jay D. Keasling. "Pathway Engineering by Designed Divergent Evolution." *Current Opinion in Chemical Biology* 11, no. 2 (April 2007): 233–39.

Zarrinpar, Ali, Sang-Hyun Park, and Wendell A. Lim. "Optimization of Specificity in a Cellular Protein Interaction Network by Negative Selection." *Nature* 426 (December 11, 2003): 676–80.

Zoloth, Laurie. "Second Life: Some Ethical Issues in Synthetic Biology and the Recapitulation of Evolution." In *The Ethics of Protocells: Moral and Social Implications of Creating Life in the Laboratory*, edited by Mark Bedau and Emily Parke. Cambridge, MA: MIT Press, 2008.

INDEX